FINITE ELEMENT SIMULATION OF EXTERNAL EAR SOUND FIELDS FOR THE OPTIMIZATION OF EARDRUM-RELATED MEASUREMENTS

Dissertation
zur Erlangung des Grades eines Doktor-Ingenieurs
der Fakultät für Elektrotechnik und Informationstechnik
an der Ruhr-Universität Bochum

vorgelegt von
Sebastian Schmidt
Bochum, 2009

Berichter:
Prof. Dr.-Ing. Herbert Hudde
Prof. Dr. rer. nat. Michael Vorländer

Tag der mündlichen Prüfung: 26. Juni 2009

Bibliografische Information der Deutschen Nationalbibliothek

Die Deutsche Nationalbibliothek verzeichnet diese Publikation in der Deutschen Nationalbibliografie; detaillierte bibliografische Daten sind im Internet über http://dnb.d-nb.de abrufbar.

ISBN 978-3-8325-2262-9

Logos Verlag Berlin GmbH
Comeniushof, Gubener Str. 47,
10243 Berlin
Tel.: +49 030 42 85 10 90
Fax: +49 030 42 85 10 92
INTERNET: http://www.logos-verlag.de

Zusammenfassung

Finite-Elemente-Simulation von Außenohrschallfeldern zur Optimierung trommelfellbezogener Messmethoden

Der menschliche Gehörgang kann grob als einseitig schallhart abgeschlossene akustische Leitung angenähert werden. Daher sind Schalldrucksignale in seinem Innern durch Resonanzspitzen und -täler geprägt. Als Bezugsgröße für audiologische und psychoakustische Messungen bietet sich der im innersten Trommelfellwinkel auftretende Schalldruck p_T an. Er weist einen bestmöglich glatten Frequenzgang auf, da sich vor dem annähernd schallharten Trommelfell bei allen Frequenzen des Hörbereichs ein Schalldruckmaximum aufbaut. Der Druck p_T hat eine wohldefinierte Position, so dass er gut reproduziert werden kann, was bei Referenzsignalen an anderen Stellen im Gehörgang nicht immer der Fall ist.

Um audiometrische Messungen auf den Trommelfellschalldruck p_T beziehen zu können, muss dieser möglichst präzise geschätzt werden, denn eine direkte Messung ist nicht praktikabel. Dazu ist es notwendig, eine Mikrofonsonde in den Gehörgang einzuführen und die Resttransformation zum Trommelfell zu berechnen, da ein im Abstand gemessenes Signal erheblich von der gesuchten Größe abweichen kann. Ein "klassischer" Ansatz besteht in der Berechnung der Gehörgangsgeometrie mittels einer Impedanzmessung, wofür gleichzeitig der Schallfluss an der Sondenposition bestimmt werden müsste. Dieses Verfahren zeigt gute Ergebnisse für inhomogene künstliche Gehörgänge mit gerader Mittelachse. Bei natürlichen, gebogenen Gehörgängen versagt die Methode jedoch häufig.

Daher wurden die räumlichen Eigenschaften des Schallfelds am Außenohr mit Hilfe von Finite-Elemente-Berechnungen untersucht. Das Modell besteht aus einer natürlichen Pinna, mehreren manuell entworfenen Gehörgängen sowie einem schwingungsfähigen Mittelohr. Die Simulationsergebnisse zeigen, dass das Trommelfell Schallwellen außerhalb der Mittelohrresonanz in den innersten Trommelfellwinkel führt. Zudem wurde gezeigt, dass der Zusammenhang zwischen p_T und Signalen, die tiefer im Mittelohr auftreten, weitgehend unabhängig von der Ausrichtung des Trommelfells ist. Der Druck p_T kann somit als Eingangssignal des Mittelohrs interpretiert werden. Innerhalb des Gehörgangs konnte eine Zone nachgewiesen werden, deren Schallfeldform nicht vom äußeren Feld abhängt, in der jedoch speziell in den Gehörgangsbiegungen deutlich irreguläre, dreidimensionale Strukturen sichtbar werden, deren Form und Position zudem frequenzabhängig ist. Diese beeinflussen zwar die Übertragungsfunktion des Gehörgangs nur wenig, führen aber bei einer Ankopplung von Messinstrumenten zu Fehlern. Durch die Simulation von Impedanzmessungen am Gehörgangseingang konnte gezeigt werden, dass oberhalb von 3-4 kHz wesentliche Fehler durch räumliche

Schallfeldeffekte entstehen, die nicht mit gängigen Netzwerkmodellen korrigiert werden können. Der Entwurf eines akustischen Gehörgangsmodells aus Impedanzdaten ist also mit entsprechenden Folgefehlern behaftet.

Diese Ergebnisse zeigen, dass es sinnvoll ist, die Übertragungsfunktion zwischen einem Messpunkt am Gehörgangseingang und dem Trommelfell direkt aus dem gemessenen Schalldrucksignal zu schätzen. Dazu wird ein physikalisches Ersatzmodell an zuvor detektierte Minima des Sondendrucks angepasst. Der Aufwand ist dabei so gering, dass eine Schätzung kurz vor einem gehörbezogenen Experiment durchgeführt werden kann. Das Verfahren wurde durch weitere Finite-Elemente-Simulationen und Messungen an einem Kunstohr evaluiert.

In mehreren Hörversuchen wurde das Verfahren in der Anwendung getestet, so dass erste experimentelle Daten mit Bezug zum Trommelfellschalldruck bestimmt werden konnten und gleichzeitig eine weitere indirekte Bestätigung der Methode gewonnen wurde. Gemessen wurden frequenzabhängige Kurven gleicher Lautstärkewahrnehmung (Isophonen). Erwartungsgemäß wird der Einfluss des Gehörgangs aus den resultierenden Kurven ausgeblendet. Zudem zeigen die Messungen wesentlich weniger interindividuelle Unterschiede als Kurven, die durch den standardisierten Bezug auf den Freifeldschalldruck die Wirkung der jeweiligen Außenohrgeometrie mit einbeziehen. Damit werden die Messdaten besser vergleichbar, interpolierbar und damit aussagekräftiger.

Table of contents

1 Introduction

1.1 Background: the sound pressure at the eardrum as a reference signal for measurements

The human peripheral auditory system converts signals from external sound sources into information input for higher organized levels of the central nervous system. By this process, a feature exctraction of the acoustical signals is obtained. Particular attributes of signals at the ear (e.g. signal amplitude, frequency, spectral composition, temporal structure) are related to those of the corresponding percept (Blauert, 1997). The mapping between the physical and the perceptive features (for instance, sound pressure and loudness) can be modeled as psychophysical function. Knowledge of such relationships helps to estimate, how physical attributes of stimuli have to be adjusted to evoke the desired perception, which is useful for many purposes (e.g. design and fitting of hearing aids, audiometry, functional tests of the hearing organ, design of devices for personal sound reproduction like headphones or insert earphones, noise reduction algorithms, compression algorithms, product sound quality judgement etc.). However, to this day, it is not possible to measure any physical signal in the organism which is exactly congruent with the auditory sensation. The percept itself is exclusively accessible to the subject in introspective experiments: listeners have to judge and describe the attributes of the auditory sensation (e.g. absolute threshold of perception or points of subjective equality for several stimuli). The psychophysical function can then be determined by relating the response of the subject to a physical reference quantity. To generalize the results of psychoacoustical investigations, data measured with various subjects must be comparable, which is only given when the physical features of the reference signal are equivalent for each subject.

In experimental setups, various signals can be selected as reference. The schematic in Fig. 1 shows essential quantities that occur when subjects are exposed to acoustical stimuli (e.g. in psychoacoustical experiments).

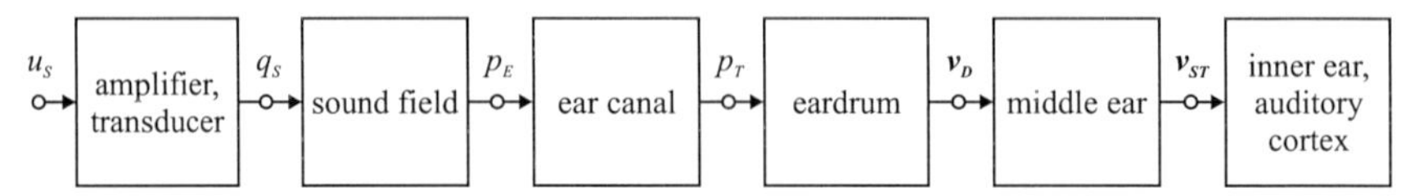

Fig. 1: Signals and systems at the peripheral hearing organ

Usually, stimuli are prepared using a digital system (PC, digital measurement device) and converted to an analog source voltage u_S. Using electromechanical transducers (e.g. earphones or loudspeaker), the signal is transformed into the source volume velocity q_S. In the adjacent air, a sound field is excited. Consequently, a sound pressure signal p_E arises at the ear canal entrance. The ear canal represents an acoustical duct. The resulting pressure at its termination (p_T) generates a force acting upon the tympanic membrane. The eardrum is loaded by the *manubrium mallei*, a part of the most anterior ossicle. Hence, the force excites velocity vibrations of the eardrum and the ossicles (v_D). The middle ear transmits the vibration to the oval window of the cochlea. At this location, the *stapes* velocity v_{ST} excites the inner ear. After frequency decomposition of the signal in the cochlea and conversion into neural impulses, the auditory nerve delivers information to the auditory cortex and higher organized levels.

It is reasonable to specify the input signal of the middle ear (represented by one of the signals occurring at the tympanic membrane) as a reference for psychoacoustical measurements. On the one hand, the influence of transfer functions between the sound source and the input to the middle ear is cancelled from the measurement results. The external ear geometry and consequently the transfer characteristics are subject to significant interindividual variations (e.g. pinna shape, cross-sectional area function, diameter and length of the ear canal, orientation of the tympanic membrane). Thus, measurements of different subjects can be compared and averaged better when they are measured with reference to the middle ear input instead of signals that do not include the influence of the entire external ear (such as earphone source voltage, free-field pressure at the position of the subject's head or pressure signals in the sound field between the source and the termination of the ear canal).

On the other hand, signals occurring at the eardrum can be accessed with reasonable effort during in-vivo experiments. This is not given for inner ear signals like the pressure in the vestibule or the vibration of the basilar membrane.

To obtain convenient measurements at the ear canal entrance, quantities that are constant in the whole external ear sound field (or at least at the reference measurement position and at the eardrum) are sometimes specified as input to the middle ear. For instance, acoustical power or intensity at the entrance of the canal and at the tympanic membrane are equivalent, when the ear canal losses are neglected. Neely and Gorga, 1998, relate hearing threshold measurements to the sound intensity at the ear canal entrance, which is calculated from the local pressure and the input impedance of the ear canal. Sound intensity is expressed as product of pressure and velocity. However, the volume velocity that is directed into the eardrum is very small, because the eardrum represents a considerably stiff membrane. Thus, it rather acts as pressure detector; consequently, it is reasonable to regard a pressure signal instead of the intensity as input to the middle ear. An approach using exclusively pressure reference quantities was introduced by Scheperle et al., 2008. A separation between the pressure waves traveling inward and outward the ear canal is performed. The magnitude of the inward traveling wave serves as measure for the acoustical input to the middle ear. However, the magnitude of pressure wave components is spatially constant only in homogeneous ducts. In generally curved and inhomogeneous ear canals, major deviations can be expected. Furthermore, both approaches are based on impedance measurements at the ear canal entrance. It is shown in this thesis that the determination of the ear canal input impedance depends on the applied measurement device and can be questionable for frequencies above 3-4 kHz.

Thus, it is favorable to relate measurements to one of the independent field variables directly. The eardrum pressure p_T can be suggested as appropriate input signal of the middle ear. Pressure is a scalar quantity that can be easily determined by a single microphone measurement, while the eardrum velocity v_D consists of three spatial components that have to be determined separately. As already mentioned, it can be assumed that the tympanic membrane serves as pressure detector, thus, auditory perception is related closely to the signal p_T. Furthermore, the sound pressure at the eardrum has essential advantages in comparison to other pressure signals arising in the ear canal. Due to the duct-like structure of the canal, distinct pressure minima which depend on the individual canal shape can be found at positions remote from the eardrum. The pressure p_T has an optimally flat frequency response, because a pressure maximum arises at the approximately rigid tympanic membrane for all frequencies of the hearing range. The position where the signal p_T occurs can be precisely specified in virtually every real ear canal. Thus, eardrum signals can be easily reproduced, which is not always possible when reference positions in the canal are selected.

1.2 State-of-the-art approaches for eardrum pressure determination

Often, the pressure that is generated inside a measurement coupler by the utilized earphones serves as a reference for hearing-related experiments (clinical and industrial applications are standardized by ISO 389-1 or ITU-T P.57, for example). Couplers consist of a cavity with standardized volume and an attached microphone (e.g. IEC711). This technique is intended to reproduce the eardrum sound pressure that arises when the earphone delivers sound to the ear canal. However, real pinnae and ear canals show a broad range of interindividually different geometries. Thus, significant deviations between the coupler prediction and the real eardrum pressure can be observed (e.g. Sachs and Burkhardt, 1972; Gilman and Dirks, 1986). To obtain the middle ear input signal accurately, it has to be determined individually.

Unfortunately, precise pressure measurements very close to the tympanic membrane are difficult. The pressure sensitivity of a microphone that is brought close to the ear canal termination changes due to proximity effects. Moreover, touching the eardrum causes unnecessary harm to the subject. In practice, the microphone has to be positioned at a certain distance from the eardrum. In several studies, the signal received by a microphone placed at an arbitrary position inside the ear canal or at a standardized distance which is controlled by optical or acoustical methods is specified as eardrum pressure (e.g. Hellstrom and Axelsson, 1993; Pralong and Carlile, 1996).

In first approximation, the pressure minimum at a quarter of the wavelength that arises in front of a rigid boundary can be used as indicator of the microphone position and to derive a simple model of the ear canal. The incident and reflected pressure waves of frequency f form a destructive interference at a distance of $c/4f$ (with c: speed of sound). Implicitly, the tympanic membrane is assumed to have a very high acoustical impedance to obtain a positive pressure wave reflection (in this context, "very high" means with respect to the ear canal tube wave impedance $Z_{twi}=\rho c/A$, with ρ: density of air and A: cross-sectional area of the canal). A method for microphone placement at the distance of the quarter-wavelength minimum is discussed in Storey and Dillon, 2001. The ear canal is excited using a sinusoidal signal at 6 kHz and the microphone position is varied until a local pressure minimum is found. This is a special case of the broadband methods listed in the following, as only one sinusoidal signal is used for the positioning process. The method is robust against pressure minima that are already present in the excitation spectrum (e.g. pinna resonances or transducer frequency response) which could be misinterpreted as standing wave minima.

When the reference microphone is placed at a distance from the eardrum, it is implicitly assumed that the local pressure does not differ considerably from the eardrum signal in the regarded frequency range. This is a reasonable approximation for positions very close to the tympanic membrane and for sufficiently low frequencies. However, studies of the ear canal sound field indicate that pressure signals measured at anterior positions in the canal may differ essentially from the eardrum pressure (e.g. Stinson et al., 1982; Gilman and Dirks, 1986; Chan and Geisler, 1990; Stinson and Daigle, 2005). The deviations are determined by the transfer function of the ear canal between the measurement position and the eardrum, which acts as inhomogeneous and curved acoustical duct (in the following, this section of the ear canal is referred to as residual canal). Many authors accentuate the need for corrections of ear canal measurements in order to determine the eardrum pressure with significant accuracy (e.g. Gilman and Dirks, 1986; Siegel, 1994; Whitehead et al., 1995; Hammershøi and Møller, 1996; Hudde et al., 1999; Stinson and Daigle, 2005; Scheperle et al., 2008).

In most approaches, features of the measured signal itself are used for the estimation of the required residual canal transfer function. In Chan and Geisler, 1990, and Siegel, 1994, the eardrum pressure is calculated by transforming signals from the measurement point to the tympanic membrane using a cylindrical model of the residual ear canal. The transformation length is estimated by evaluation of the quarter-wavelength minimum in the sound pressure spectrum. At high frequencies, however, significant errors are observed. Chan and Geisler, 1990, show that the accuracy of the microphone-to-eardrum distance estimation depends on the curvature of ear canals (the results are shown for a replica of a natural ear canal). Ear canal shapes differ considerably from simple cylindrical models. The cross-sectional area function of the ear canal has significant influence on the transformation accuracy. Higher order minima are not compensated correctly, as the corresponding minimal frequencies that are determined by the inhomogeneous shape of the canal generally do not match the minima that are predicted by the cylindrical model.

Stevens et al, 1987, estimated the ear canal transfer characteristics using an all-pole filter that was adapted to pressure minima in an acoustical tube attached to the ear canal. Due to the relatively large distance between the microphone and the ear canal termination, numerous zeros are found in the considered frequency range and have to be compensated. The large transformation distance must be assumed as main drawback of this method. Furthermore, higher order mode effects at the coupling surface between the tube and the ear canal can decrease the accuracy of this method systematically.

When the ear canal is considered as inhomogeneous acoustical duct, its input impedance is determined by the cross-sectional area function and its termination. Consequently, the canal shape can be estimated from impedance measurements at the ear canal entrance by inverse procedures, such as methods developed for geometry estimation of the vocal tract (Schroeder, 1967; Sondhi and Gopinath, 1971; example methods for ear canal impedance measurements are listed in section 2.5). When the cross-sectional area function of the canal is known, the transfer function between a measurement point at the canal entrance and the eardrum can be calculated. This approach was implemented by Hudde et al., 1999 and evaluated in artificial inhomogeneous, but straight ear canals. Although it worked well in the used canal replicas, the method often failed in real ear canal geometries that exhibit individual curvature. This problem implies that the determined impedances and transfer functions are affected by spatial sound field effects that are not sufficiently included into the traditional modeling approaches.

1.3 Aim of this work

The main goal of this work is to provide a precise, robust and efficient method for the estimation of the sound pressure at the eardrum. For the development of the estimation technique, fundamental research concerning the external ear sound field is necessary. The approaches mentioned in section 1.2 are based on one-dimensional modeling of the ear canal, i.e. it is assumed that pressure waves predominantly travel as fundamental mode. As the inconsistent results of Hudde et al., 1999 suggest, three-dimensional sound field structures may essentially influence the accuracy of one-dimensional model calculations. Thus, a thorough analysis of the spatial sound field structure at the external ear is necessary. In this work, the field structure is obtained by numerical simulations using the Finite Element (FE) method, which has significant advantages compared to measurements. The sound field is not affected by the presence of measurement equipment. In contrast, the impact of microphone probes or impedance measurement tubes can be analyzed systematically by adding such devices. Further, the geometry of the external ear can be easily controlled by CAD design. Using the FE model, it is checked if the necessary conditions for the application of simplified one-dimensional models are fulfilled. As tympanic membrane and middle ear are simulated as well, it can be shown that p_T is well-suited as input to the middle ear. Regarding the development of the estimation method, it has to be investigated, how transfer functions in the ear canal and impedance measurements at the ear canal entrance depend on the three-dimensional sound field features. Based on the findings gained from the model, a precise and feasible method for the estimation of p_T is to be developed. The technique is evaluated and applied to psychometric measurements in a preliminary experimental study.

In Chapter 2, the FE model development and simulation results concerning the structure of the external ear sound field and measurements at the ear canal entrance are documented. First, a short overview on the anatomy of the external ear (section 2.1) and on one-dimensional modeling concepts (section 2.2) is given. After the FE model is presented in section 2.3, the spatial structure of the external ear sound field is examined in detail (section 2.4). In particular, the regions at the eardrum and at the canal entrance are analyzed. To cope with the three-dimensional shape of the wavefronts in the canal, the concept of "fundamental sound fields" in ducts is introduced. The consequences of the three-dimensional sound field for measurements at the ear canal entrance are evaluated in section 2.5.

Chapter 3 describes the developed estimation method for the eardrum sound pressure using a single pressure probe microphone. Details on the estimation process are discussed in section 3.1. The method was evaluated using further FE models and measurements in an artificial ear (section 3.2). The results of the pilot study (equal-loudness level contour determination with reference to the eardrum sound pressure) are presented in section 3.3.

Chapter 4 concludes this thesis with a summary and proposals for future work.

2 The external ear sound field

In this chapter, the characteristics of the sound field in the ear canal are examined in detail. After a short description of the external ear anatomy, approaches for one-dimensional modeling of the external ear are summarized. Such methods provide basic insight into the functionality of the external ear; however, the spatial pressure distribution inside the canal is not simulated. An accurate three-dimensional mapping of the sound field is obtained by the finite element (FE) model of the external ear that is introduced afterwards. The model answers the following questions:

- How does the distinctly three-dimensional external ear field merge with the ear canal sound field?
- Is it possible to specify a central region that is independent from the external field and the higher order modes generated by the eardrum?
- Does the sound propagation in the central region of the canal obtain regular structures, which can be expected in duct-like geometries? How does the cana curvature influence sound propagation?
- Does the eardrum guide waves to the termination point or are the waves directed into the eardrum?
- Is it possible to specify the eardrum signal in natural ear canal geometries?
- Is it possible to estimate the eardrum pressure from the canal entrance?

After an analysis of the coupling of the external sound field and the waves in the ear canal, three-dimensional structures that particularly arise in the canal bendings, "one-sided isosurfaces", are described. These shapes cannot be represented by simplified one-dimensional models in principle. The field near such regions, however, transmits sound waves with minimal energy density, a criterion that is fulfilled for fundamental mode wave propagation as well. Thus, the concept of a "fundamental ear canal sound field" is introduced. It is shown that the influence of one-sided isosurfaces on the ear canal transfer functions in the fundamental sound field is minimized due to the minimal energy density, although distinct effects of evanescent higher order modes are visible. For precise estimations of the eardrum pressure, measurements at the ear canal entrance are necessary. Before an estimation method can be developed, it is required to examine, how three-dimensional sound field effects affect measurements. Hence, this chapter is concluded by an analysis of errors that arise when the ear canal entrance impedance is determined.

2.1 The human external ear: anatomy and function

The external ear determines the transfer of sound from the field surrounding a listener to his tympanic membrane. It consists of pinna, ear canal (*meatus*) and eardrum or tympanic membrane. In a wider sense, reflections at parts of the body (e.g. the torso or the shoulders) contribute to the external ear acoustics as well. All effects superpose and form the sound field near the pinna.

The pinna is a cartilage structure that is covered with fat tissue and skin. Shape and size of individual pinnae show great diversity. The deep cave where the transition between the pinna and the ear canal can be found is called *cavum conchae*. The pinna connects the external field with the the ear canal. In the acoustical radiation impedance seen from the ear canal outwards, several peaks and notches can be observed that are originated by resonances in the folded pinna structure. As shown in section 2.4, the resonances originate from successive reflections between nearly parallel structures of the tissue. As the excitation of the resonances depends on the incidence direction of sound waves, the pinna acts as directional filter, at least for frequencies above 4 kHz.

From the *cavum*, sound is transmitted into the ear canal. It is difficult to specify the entrance of the ear canal exactly, because it tapers and merges with the skin tissue in the *cavum conchae*. Shape and length of human ear canals show considerable interindividual variations. Stinson and Lawton, 1989, compared 15 canals which yield lengths between 27 and 35 mm and average diameters between 6.2 and 8.4 mm. In each of the examined canals, two distinct bendings are noticeable. The first curve is located directly behind the *cavum conchae*, the other is found approximately halfway between the first curve and the tympanic membrane.

Otherwise, ear canals show irregular cross-sectional area shape. The canal walls are covered with skin, which merges smoothly with the tympanic membrane. Approximately two thirds of its anterior length are surrounded by cartilage and soft tissue which are attached to the pinna. The posterior third of the canal is enclosed in the nearly rigid temporal bone.

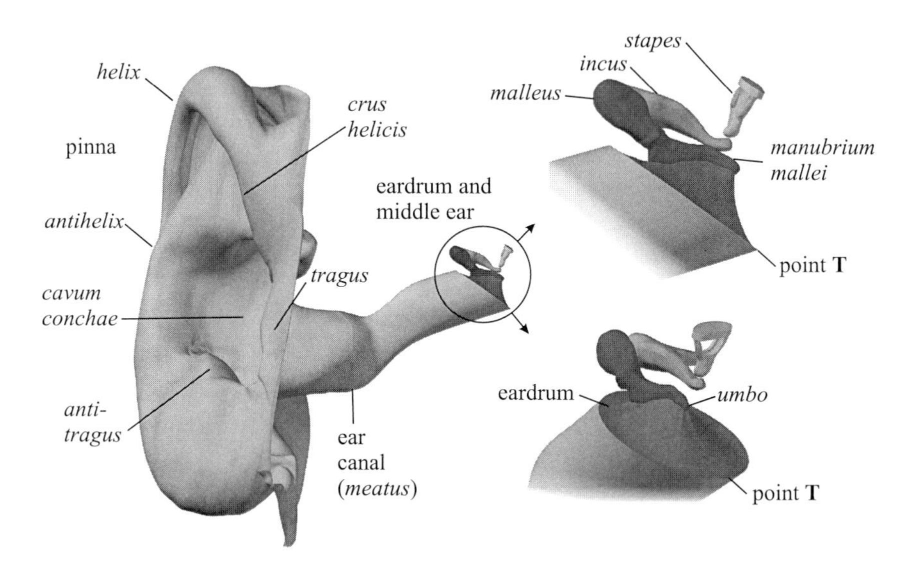

*Fig. 2: Anatomy of the human external and middle ear. Left: pinna with ear canal and ear-drum, right: two detail plots of eardrum and middle ear. The ear canal termination point **T** will be explained in section 2.2. The figures are taken from the CAD model introduced in section 2.3.*

The eardrum or tympanic membrane consists of very inhomogeneous skin tissue which is distinctly structured with fibers. It is inclined by approximately 45 degrees to the middle axis of the ear canal. Along the rim, it is clamped to the surrounding temporal bone. The tympanic membrane has roughly conical shape, as it is directly connected to the *malleus*, one of the middle ear bones. The conical apex of the tympanic membrane, the *umbo*, is slightly displaced from its center. Two zones with different stress can be identified. In the lower part, near the innermost point of the ear canal (tympanomeatal corner or point **T**), the tympanic membrane is rather tense, whereas it is flaccid in the larger upper part (*pars tensa* and *pars flaccida*). The eardrum obtains complicated vibrational patterns. Displacement measurements using optical methods show that the *malleus* divides it into two vibrating parts (Tonndorf and Khanna, 1972). As a membrane, the tympanic membrane transforms the local pressure into a force and thus excites the ossicle chain, which represents its mechanical load. The latter consists of three small bones (*malleus*, *incus* and *stapes*), which are joined to each other and to the surrounding tympanic cavity by ligaments and tendons. The ossicle chain can be prestressed by two muscles (*m. tensor tympani* and *m. stapedius*) to reduce the transfer factor of the middle ear at high eardrum pressure levels. The air volume enclosed in the tympanic cavity and the adjacent mastoid air cells acts as additional acoustical spring load on the tympanic membrane and has a stiffening effect. However, as its impedance is small against the ossicle load, it is neglected in the following. Fig. 2 shows the components of the external and middle ear. The figures are taken from the CAD model that will be introduced in section 2.3.

2.2 One-dimensional modeling of the external ear

Ear canals can be roughly approximated as narrow acoustical ducts. An acoustical duct is an enclosed fluid region in which sound waves propagate along a direction that can be uniquely specified. Sound fields in special ducts can be decomposed into different wave propagation modes. An analytical specification of modes is possible exclusively in structures that can be geometrically expressed in one of 11 specific coordinate systems (Evans, 1990). In these cases, the solution of the spatial sound wave equation that is expressed as function of one of the field variables (sound pressure p or volume velocity q) can be partially separated for one dimension (see Appendix A.1). The fundamental mode sound field can be separated into an incident and reflected wave with identical shape. The fundamental mode in a cylindrical duct, for instance, is represented by planar wave propagation parallel to the middle axis. Hence, sound pressure and velocity are constant in cross-sectional areas normal to the middle axis of the tube. Waves propagating in fundamental mode shape obtain minimal field energy density. Higher order modes in circular cylinders include radial or azimuthal wave patterns. Generally, the total sound field energy density rises, when higher order modes occur.

Higher order modes can propagate above a particular cut-off frequency only. Thus, for low-frequency approximations of the sound field in acoustical ducts, higher order modes can be neglected. In irregularly shaped ducts that do not conform with one of the suitable coordinate systems, no exact analytical representation of modes can be specified (this leads to the definition of a fundamental ear canal sound field). However, often approximate fundamental and higher order modes can be assigned to the field shape. For simplification and approximative modeling, it is reasonable to assume mainly fundamental mode propagation in ear canals, as the transversal dimensions usually are significantly smaller than the acoustical wavelengths occurring in the hearing frequency range.

The sound field in axially symmetric acoustical ducts with smoothly varying cross-sectional area function is modeled by the well-known Webster approximation (Webster, 1919). It can be derived using a generalized continuity equation. The expression represents the acoustical wave equation supplemented with a term incorporating the cross-sectional area function $A(x)$:

$$\frac{\partial^2 p}{\partial x^2}+\frac{1}{A(x)}\frac{dA(x)}{dx}\frac{\partial p}{\partial x}=\frac{1}{c^2}\frac{\partial^2 p}{\partial t^2} \tag{2.2.1}$$

The variable c represents the speed of sound. The solution of the Webster equation depends on one spatial coordinate only, which is specified along the straight middle axis x of the model. As the Webster equation implies planar fundamental mode wavess, exclusively ducts with smoothly varying cross-sectional area are modeled without significant error. Otherwise, deviations due to higher order modes excited at area discontinuities arise.

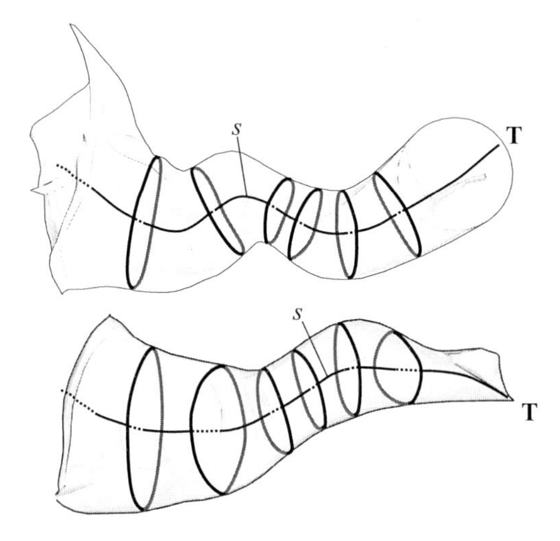

*Fig. 3: Possible middle axis **s** with corresponding orthogonal surfaces. The middle axis ends in the termination point **T** in the tympanomeatal corner. The ear canal is shown from top and front view. The pinna is located at the left hand side of the figure.*

The curvature (or in general, the three-dimensional shape) of the duct is not taken into account. Real ear canals, however, feature significant curvature. Thus, it is not possible to develop one-dimensional Webster models as strictly exact equivalents of natural ear canals. Stinson and Khanna, 1989, utilize the mass centroids of cross-sectional canal areas to interpolate a middle axis and propose a generalized formulation of Webster's equation that depends on the curvilinear middle axis *s*.

The area function *A(s)* is defined as function of the arc length along the *s*-axis. Models based on this concept are often referred to as "unidimensional" instead of "one-dimensional", as only one parameter is utilized, although the curvilinear middle axis is specified in three-dimensional space. A model of a curved middle axis with orthogonal surfaces is shown in Fig. 3. Although an adequate definition of the middle axis has been accomplished, the distinctly three-dimensional shape of the sound field due to the curvature of the canal is not modeled by the mentioned approach. The underlying sound field model assumes planar wave propagation areas perpendicular to the middle axis *s*. A modified horn equation that involves fundamental mode propagation with curved equipotential surfaces was introduced by Agullo et al, 1998. The theory was developed for axisymmetric horns featuring a straight middle axis. Farmer-Fedor and Rabbitt, 2002 applied the results to curved ducts. Thus, a unidimensional wave propagation theory including curvature and general cross-sectional area shape became available.

The mentioned approaches provide one-dimensional models for that region of the ear canal in which the sound field is independent from the spatial external field and from the near field of the eardrum. This section of the ear canal is often referred to as "core region" (Farmer-Fedor and Rabbitt, 2002). In the core region, constant-pressure surfaces are assumed which can be interpreted as ports of the one-dimensional model. The corresponding volume velocities can be calculated as intergral of the particle velocities over the port surfaces. Depending on the modeling approach, port surfaces are planar or slightly curved. Shape and position of the surfaces do neither depend on frequency nor on the external, three-dimensional field per definition. When exclusively linear transmission is assumed, the entire core region can be modeled as two-port element which transforms pressure and volume velocity according to its equivalent chain matrix (see Appendix A.1. The complex phasors of the field quantities in frequency domain are indicated by underlines in the following).

In Hudde and Engel, 1998a, a network approach was proposed which is reproduced here slightly modified. The model topology is given in Fig. 4. The pressure signal that is formed at the port surface **E** by the external field can be expressed as equivalent Thevenin source consisting of the ideal pressure source $\underline{p}_{ecE}$ and the internal impedance $\underline{Z}_{ecE}$. Both elements are specified relative to the entrance port **E**, which is element of the core region. Thus, shape and position of the corresponding constant-pressure surface are independent of the source. The signal $\underline{p}_{ecE}$ represents the pressure that is arises on the surface **E** for a given external sound field, when the ear canal is blocked locally by a rigid wall. The impedance $\underline{Z}_{ecE}$ is identical to the radiation impedance seen outwards from the port **E**. With the mentioned one-dimensional fundamental mode approaches, the core region of the ear canal between the ports **E** and **X** is represented by a two-port $\underline{C}_{EX}$. If the canal shape is known, the appropriate chain matrix can be calculated from a generic or modified Webster equation.

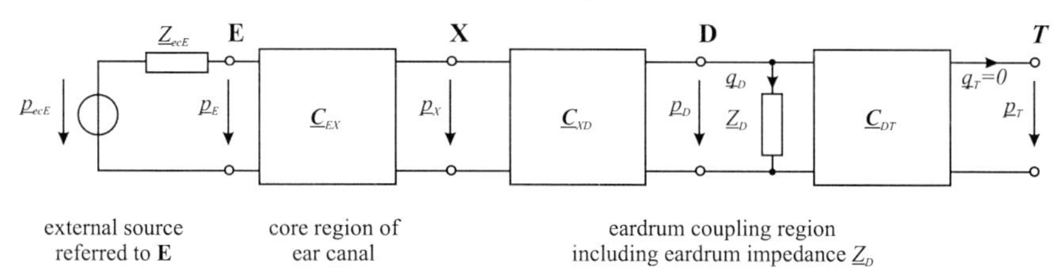

Fig. 4: Modified topology of the one-dimensional external ear model proposed in Hudde and Engel, 1998a.

The sound pressure near the eardrum cannot be predicted exactly using models that incorporate exclusively fundamental modes. Due to the eardrum shape and its vibrations, a local sound field composed of evanescent higher order modes is generated. As simplification, the circuit model implements sound waves that are basically guided by the tympanic membrane as if it were rigid. Thus, waves propagate along a continued middle axis into the innermost corner between the eardrum and the canal wall (tympanomeatal corner). The point **T** represents the rigid termination of the model, where the volume velocity $\underline{q}_T$ vanishes. Incident sound pressure (and power) is completely reflected in **T**. To account for the energy dissipation of the tympanic membrane, a lumped eardrum impedance element $\underline{Z}_D$ is inserted between the two-ports $\underline{C}_{XD}$ and $\underline{C}_{DT}$. The corresponding port area **D** is represented by the equipotential surface running through the *umbo*. The volume velocity $\underline{q}_D$ that appears in the model is directed into the normal direction of the tympanic membrane. The section of the ear canal that is modeled by the three elements $\underline{C}_{XD}$, $\underline{C}_{DT}$ and $\underline{Z}_D$ is referred to as "eardrum coupling region" in the following.

The mentioned extensions of Webster's equations are based on fundamental mode propagation. Thus, they represent low-frequency approximations, because higher order mode propagation is exclusively possible for frequencies above a particular cutoff value. However, sound fields can be affected by evanescent higher order modes even for frequencies that are distinctly lower than the cut-off limit. A one-dimensional model that takes higher order modes into account was introduced by Rabbitt and Holmes, 1988 for axisymmetric ducts with a straight middle axis. A central issue of this study was the investigation of higher modes at the coupling of the distinctly three-dimensional sound fields at the pinna and the eardrum. Later, the method was extended for general cross-sectional area functions (Rabbitt and Friedrich, 1991). A comparison between the predictions of analytic horn equation approaches and a numerical sound field calculation is already documented in Stinson and Daigle, 2005. Along the middle axis, the one-dimensional algorithm determined the pressure sound field fairly well. However, in planar surfaces perpendicular to the middle axis, the pressure varied significantly. In conclusion, ear canal modeling approaches that exclusively imply fundamental mode propagation cannot correctly reproduce the spatial canal sound field, even within the core region.

2.3 Modeling with finite elements

For the issues that are investigated in this thesis, no new analytical method is required which could further improve the computation for individual canal geometries. Rather, the attributes of the field and the limits of unidimensional approaches using fundamental modes are of interest. In particular, the three-dimensional structures which arise in natural ear canals due to bendings and curvature have to be examined. Thus, the sound field has to be mapped spatially. As straightforward approach, the fields that arise in natural ear canals or replicas can be scanned using thin probe microphones. However, such measurements can be very inaccurate. It is not easy to locate the tip of the microphone with significant precision. A large number of successive measurements would be necessary to obtain a sufficiently dense data grid. The microphone may introduce sound field disturbances that cannot be separated easily from the effects of the ear canal curvature; hence, it is necessary to use very thin probes (e.g. probes with diameter of 0.2 mm as used in Stinson and Daigle, 2007). However, extremely thin microphone probes may exhibit poor signal-to-noise ratio.

An accurate and efficient representation of sound fields that avoids the presence of measurement devices can be achieved by numerical models. It is another main advantage of such methods that the boundary geometry can be efficiently controlled by CAD methods. Generally, the solution of the acoustical wave equation under the given boundary conditions (e.g. ear canal walls, elastic tympanic membrane, sound source etc.) has to be found. In an earlier study, finite differences were applied to calculate time domain pressure signals at the external ear (Schmidt and Hudde, 2004). Regarding the structure of the acoustical wave equation (see Appendix A.1), the time domain formulation of finite differences is a straightforward approach for sound field simulations (finite difference time domain or FDTD, Botteldooren, 1994). In boundary element (BE) simulations, the sound pressure distribution on the surface area of the simulation region is calculated. In acoustics, the solution is based on Stokes' theorem which reduces volume integrals to surface integrals. The field inside the volume can be calculated by solution of the Kirchhoff-Helmholtz integral equation. The surface is represented as a set of discrete elements. Finite element (FE) simulations use a spatial discretization instead of the surface integral approach. Hence, during the solution of the model equation system, all field values are calculated directly and no post-processing is necessary. However, the sum of elements for spatial discretization of a volume is generally larger than the amount of surface elements.

Each of the three simulation methods can be applied either in time or in frequency domain. The evaluation of time domain results in frequency domain is exclusively possible, when the stationary oscillation state of the modeled system is reached within the simulation time. Hence, time domain models are best suited for systems with short impulse responses. However, this condition is not fulfilled for ear canals that show strong resonances. For the required sound field calculations, finite element simulations in frequency domain were implemented, because important "modules" of the human auditory system were already available as FE model (Weistenhöfer and Hudde, 1999; Weistenhöfer 2002; Curdes *et al.*, 2004; Taschke, 2005; Taschke and Hudde, 2006) and could be added to those parts of the models that were designed exclusively for the external ear studies. Furthermore, a large range of software tools for geometry acquisition and data pre- and post-processing was accessible from previous projects. For the current study, the software package was extended with programs providing efficient geometry intergrity checks and data exchange.

In this work, the commercial finite element software ANSYS® was used. For the formulation and solution of an FE calculation, a digitized representation of the simulation region is necessary. In the context of ANSYS®, the volumes that form the simulation region are composed of its boundary areas. Areas are modeled by adjacent lines, which are in turn defined by spline data. Initially, the geometry objects (volumes, areas, splines and keypoints) are imported from CAD designs. For spline drawing and volume merging, AUTOCAD® was used. The meshing algorithm of ANSYS® automatically divides the simulation region into hexahedral or tetrahedral subregions.

A thorough discussion of the finite element method is available in standard textbooks (e.g. Petyt, 1990; Bathe, 1996; Hughes, 2000; Zienkiewicz and Taylor, 2000). The implemented models of ear canals are very simple acoustical systems. Wave effects are perfectly reproduced in FE simulations, especially when propagation or boundary losses are not taken into account. It is, however, necessary to use a sufficiently fine FE mesh. Throughout this work, the element edge length was iteratively determined by refining the mesh until the simulation results did not depend on the spatial discretization any more.

The models are based on some simplifying assumptions. It was already mentioned that harmonic (frequency domain) analyses are carried out. Thus, the models are not capable of simulating nonlinear effects (e.g. nonlinear tympanic membrane vibrations). However, it can be expected that nonlinear characteristics can be neglected in comparison to the linear distortions caused by the external ear transfer functions. Furthermore, propagation losses are not included into the model, although the damping inside the ear canal is noticeable in measurements. However, for the basic research issues that are discussed in this thesis, the influence of the canal shape and the eardrum vibration and energy absorption are significantly more relevant.

The model geometry was assembled from variably generated components. It was not necessary to reproduce any particular external ear anatomy precisely, because only the essential features of the sound field arising in a plausible geometry had to be investigated. For instance, different ear canals were designed manually considering the shape of replicas taken from natural ear canals (Stinson and Lawton, 1989). In contrast, the complex pinna geometry was extracted from anatomical imaging scans of a human subject.

2.3.1 Pinna

The major modeling effort was spent on the pinna, as it essentially determines, how the external spatial sound field merges with the duct-like field in the ear canal. Natural pinnae are structured significantly and have complicated shape. Resonances in the grooves of the pinna influence its acoustical transfer features essentially. These characteristics should be obtained in the simulations. As it is not a trivial task to generate a CAD pinna model from scratch, semiautomatic acquisition of a natural pinna geometry was carried out. Thus, a human pinna was scanned by magnetic resonance imaging (MRI) and a contour detection tool was developed to extract splines from the MRI contours using MATLAB® (see Fig. 5).

First, a layer of the MRI data set is selected, loaded and displayed on the main panel. Contrast and brightness of the image are adjusted to optimize its signal-to-noise ratio. In the given context, the brightness of the tissue image is increased and the visibility of artifacts and noise is decreased. Additionally, a nonlinear (logarithmic) transformation and a pixel value threshold which masks artifacts can be applied to the image data to obtain a maximum contrast the pinna tissue and the surrounding volume. To improve the quality of edges, several standard image processing filters are available (mean, median and lowpass filter). For edge detection, built-in MATLAB® functions were used which provide several detection algorithms (gradient methods using Sobel, Prewitt or Roberts approximation, zero crossing methods and gradient-threshold techniques like the Canny method; Parker, 1997) which can be chosen by the user.

The contour detection is refreshed after each image processing step and can be displayed simultaneously with the original image. The processing steps are recorded as "stack". If a set of steps has sufficiently good detection results, its stack can be saved and applied to other cross-sectional layer images. Simulataneously, the contours are converted into splines which can also be displayed in the main image panel. If a reasonable approximation of the underlying image is found, the splines are saved. Using a further tool, the set of splines can be optimized by reducing the number of sampling points and smoothing the contours.

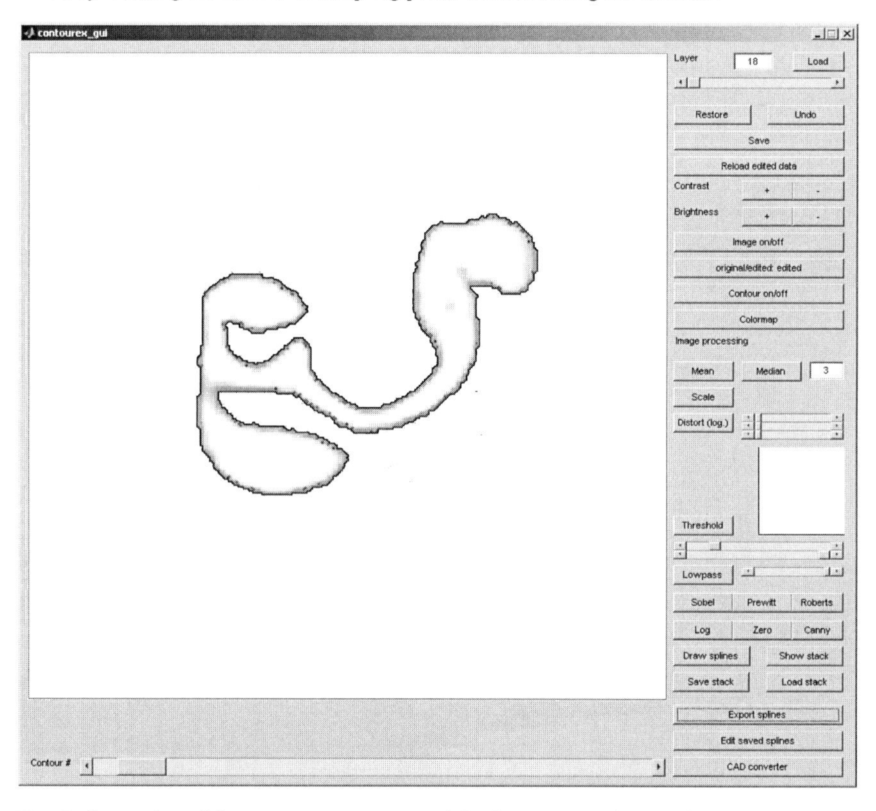

Fig. 5: Screenshot of the contour extraction tool. In the main panel, one of the cross-sectional pinna slices used in the modeling process is displayed. The image is already processed and a spline contour (gray color) is drawn around the selected region.

After that, the spline data is used to generate an AutoLISP code automatically from the MATLAB® routine. The resulting routines are applied to control AUTOCAD® using batch processing scripts. When the script is executed, level curves of the pinna are drawn by AUTO-CAD® as splines. The lateral coordinate is calculated from the position of the respective MRI cross-section. The necessary orthogonal splines were added manually and the geometry was optimized for FE purposes (e.g. prevention of extremely sharp angles). The resulting pinna model is shown in Fig. 6 (and already in Fig. 2). Some marks of the original splines can be found as vertical parallel lines of the pinna in the top right panel of Fig. 7.

The surface areas of the pinna are modeled as rigid boundary of the external air volume in the simulation. This is a reasonable approximation regarding the tissue surfaces at the head, as studies of the local acoustical impedance (Katz, 2000) show. Sound radiation from pinna vibrations can be neglected as well, in particular regarding the aim of the present work. Hence, the tissue is not modeled. In the FE model, the pinna represents a totally rigid and clamped solid object.

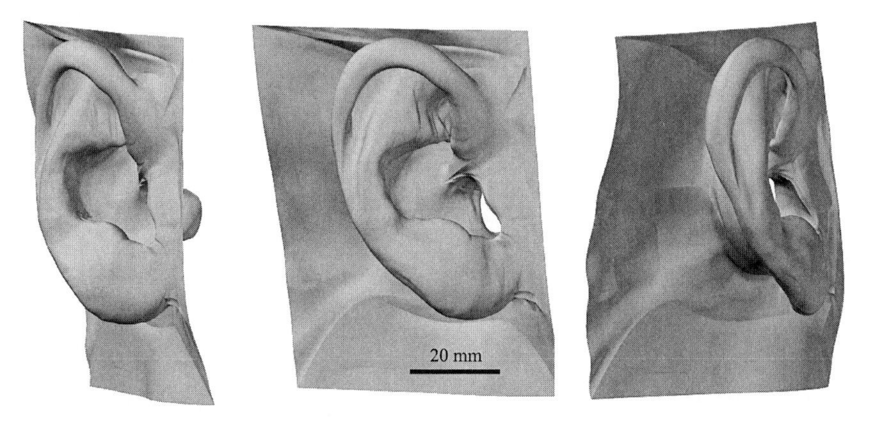

Fig. 6: Pinna model obtained from MRI scans of a human subject seen under three different angles.

2.3.2 External air volume

The pinna has to be excited by sound waves from an external volume filled with air. On the one hand, the volume must be large enough so that numerical or acoustical effects originated by the boundaries do not disturb the sound field arising at the pinna structures. For instance, this allows for correct calculations of the pinna radiation impedance. Further, sound sources have to be placed at a sufficient distance from the pinna to avoid near field interactions.

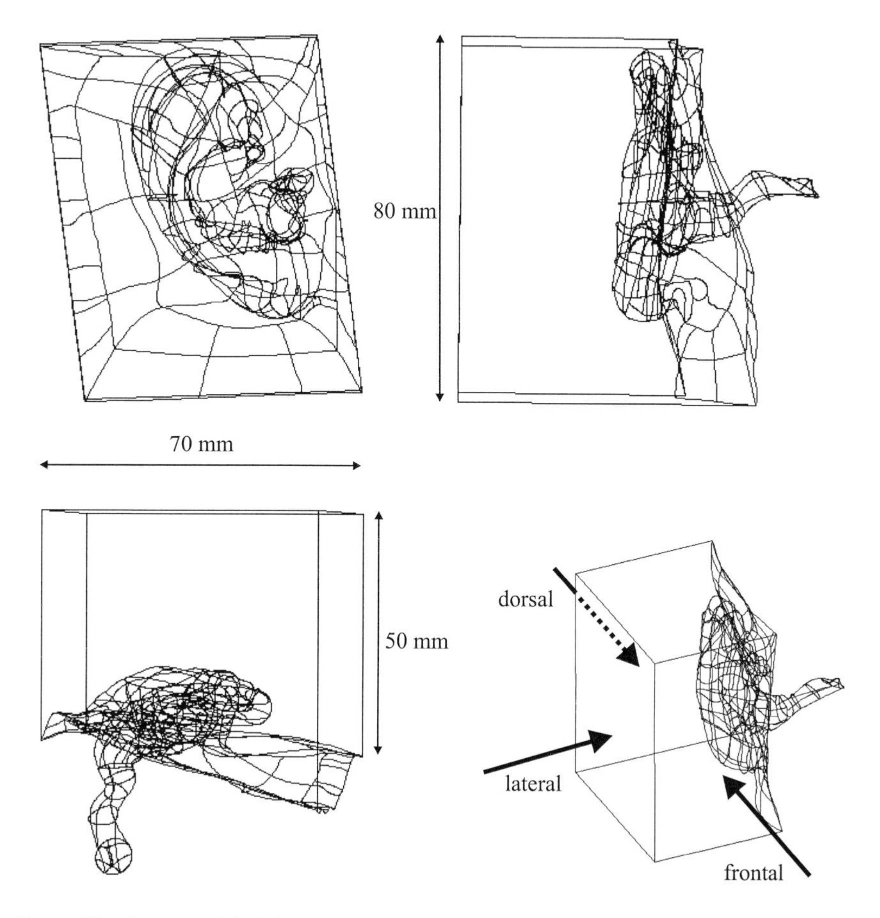

Fig. 7: Wireframe model of the external air volume seen from orthogonal directions (top row, bottom left) and directions of excitation (bottom right).

On the other hand, the computation expense of the model is essentially determined by the size of the external volume, because it is the largest element of the simulation region. Thus, the volume should be kept small to reduce the calculation time. It is not possible to simulate widespread sound fields around the head efficiently using an FE model. Far-field studies such as HRTF simulation are best carried out with boundary element models (Katz, 2001; Fels, 2008). It turned out that a relatively small volume is sufficient for the aims of this study. For easy geometry handling, it was implemented with almost rectangular boundaries. In the following, it is referred to as "pinna box". The volume was meshed with an average nodal distance of 3 mm (approximately 7 elements per wavelength at 16 kHz).

The boundary areas serve as sound source for the model calculations. All nodes on the specified boundary are excited with constant displacement magnitude and phase angle. Thus, vibrating pistons (instead of membranes) are simulated. The area vibrations are coupled to the field by special elements that obtain fluid-structure-interaction. The displacement source excites a volume velocity field ($\underline{q}_0 = \underline{v}_0 A = j\omega\xi_0 A$). To avoid strong reflections at the source surface, it is covered with absorbing elements. For the examination of the source influence on the ear canal sound field, three distinctly different excitation configurations were realized (frontal, dorsal and lateral source; see Fig. 7). Although the sources are comparably close to the pinna, grazing incidence of sound waves from frontal and dorsal directions or impinging incidence from the lateral source is approximated, as can be seen later in the results (Fig. 21).

Sound waves directed outward should be absorbed at the boundaries of the region. Anechoic conditions are approximated by implementing a boundary condition on the walls that models the specific field impedance $Z=\rho c$. The variable ρ denotes the static mass density of air. Such surfaces absorb the normal components of incident sound waves totally.

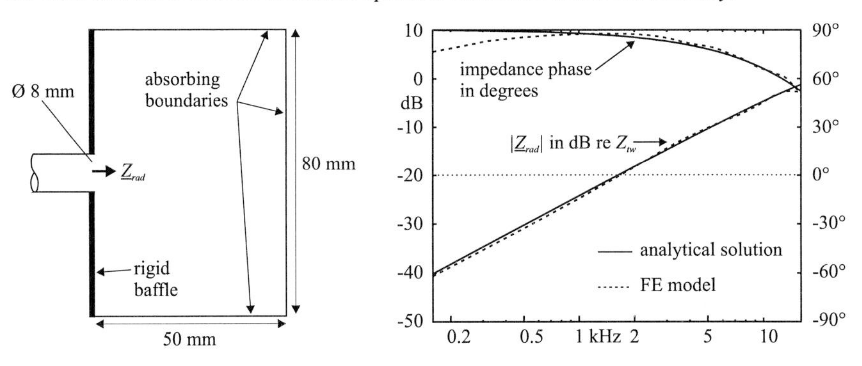

Fig. 8: Left: cross-sectional view of the simplified system for a preliminary test of the pinna box concept. Right: Logarithmic magnitude of the simulated radiation impedance (solid line) and the analytical solution (dashed line). Both impedances are normalized by the tube wave impedance of the test duct.

To test whether the pinna box is a feasible approximation of a volume with totally absorbing boundaries, a simplified acoustical system with analytically known radiation impedance was simulated. The size of the model air volume is chosen similar to the pinna box. In the model, an acoustical duct is mounted flush in the center of a baffle (see Fig. 8, left panel). The acoustical radiation impedance seen at the orifice of a tube mounted in an infinitely large baffle can be expressed by the term (Zollner and Zwicker, 1993)

$$\underline{Z}_{rad} = \frac{\underline{p}}{\underline{q}} = \frac{\rho c}{A}\left(1 - \frac{J_1(2\beta r)}{\beta r} + j\frac{H_1(2\beta r)}{\beta r}\right) \tag{2.3.1}$$

The tube has cross-sectional area A and radius r. The functions J_1 and H_1 denote Bessel and Struve functions of the first order. From the test model, the same impedance can be calculated by the procedure described in subsection 2.4.1.

The results are displayed in Fig. 8, right panel. Obviously, the pinna box is a good approximation for free space around the pinna. For frequencies less than 1 kHz, the impedance phase angle is underestimated. The deviation is most likely caused by the small box volume, however, it can be considered as minor effect.

2.3.3 Ear canals

Although the shape of natural ear canals is almost arbitrary (e.g. Stinson and Lawton, 1989), some common features exist. Ear canals are tapering ducts with significant curvature. In most canals, two bendings can be distinguished. The shape of cross-sectional areas changes over the length of the canal. For the FE model, several ear canals were designed manually which comply with these features.

The mentioned canal replicas by Stinson and Lawton served as guideline, when the canal models were matched to the pinna and eardrum areas. First, the approximate length of the canal was set by adjusting the distance between the pinna and the tympanic membrane model. After that, auxiliary lines were drawn between the rim of the eardrum and various points on the pinna surface. Using the auxiliary lines, four master splines were constructed and deformed manually if necessary.

The splines were merged flush with the surface areas at the *cavum conchae* of the pinna. Additionally, several closed splines were drawn around the bunch of master splines. Finally, areas were constructed in the spline grid. The resulting ear canal volumes were transferred to the FE system and meshed with fluid elements. The average distance between nodes in the ear canal was adjusted to 1.5 mm (approximately 15 elements per wavelength at 16 kHz). Like the surface areas of the pinna, the walls of the ear canal model are rigid and clamped.

After the basic characteristics of the external ear sound field were examined using one canal model (Fig. 9), an additional set of canals was designed to test the influence of the canal shape on the eardrum pressure estimation algorithm. These canals are displayed in Fig. A.5.2.

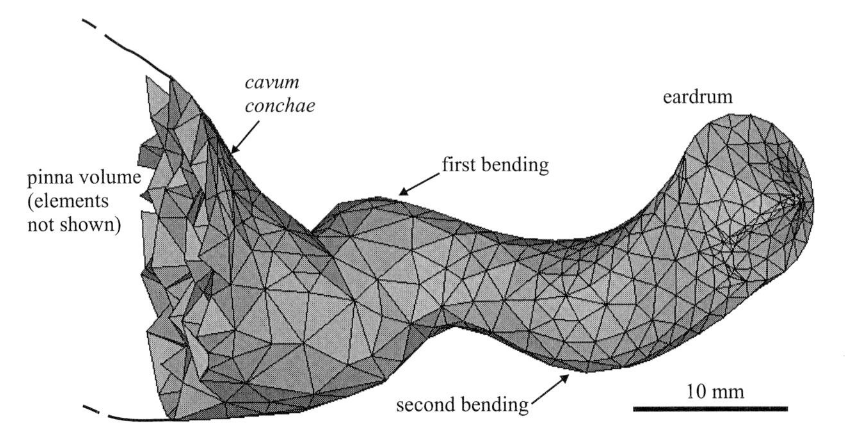

Fig. 9: Meshed model of the ear canal used for the sound field examinations. The canal obtains two significant bendings and an arbitrary cross-sectional area shape.

2.3.4 Tympanic membrane and middle ear

The eardrum can be expected to evoke higher order modes at the end of the ear canal (Rabbitt and Holmes, 1988). These are originated by the retracted conical shape of the tympanic membrane on the one hand and by its vibrations on the other hand. The eardrum is a very thin and stiff tissue membrane. It is simulated by two-dimensional shell elements. The membrane shape is reproduced from a model by Weistenhöfer, 2002. Alike natural tympanic membranes, the model eardrum is divided into a flaccid and a tense region. The respective elements have constant mass density and Poisson's ratio, but different elastic moduli.

As the vibrational characteristics of the eardrum are essentially determined by its mechanical load, a complete middle ear model is implemented. The shape of the middle ear bones was measured by Weistenhöfer using an optical method (Weistenhöfer and Hudde, 1999). In the original middle ear model, the joints and tendons linking the ossicle bones were designed as lumped elements referring to an externally stored generalized impedance matrix. As this method restricts the use of the model significantly, the generalized impedance elements were replaced by simplified equivalents formed as circular cylinders and meshed using solid elements. The geometry (radius and length) and material parameters (mass density, Poisson's ratio, elastic modulus and damping) were adapted in such a way that transfer functions and impedances of the middle ear were consistent with the previous model. The middle ear is suspended by ligaments that are connected to the walls of the tympanic cavity. The model ligaments are clamped at these points. The air inside the tympanic cavity is not included in the model, as the major load on the tympanic membrane is applied by the ossicles. The effect of the tympanic cavity is modeled by the equivalent circuit in Fig. 10.

The circuit on the left represents the eardrum which converts the local pressure into a mechanical vibration (first gyrator). The major fraction of the force acts upon the lumped mechanical impedance element $\underline{Z}_{mech}$ which incorporates both the ossicle load and the impedance of the eardrum itself. The remaining force excites a pressure in the tympanic cavity which is modeled as acoustical compliance n_{TC}. The equivalent network can be simplified to evaluate the acoustical load on the ear canal. The results show, that the omitted air volume would have a stiffening effect on the eardrum.

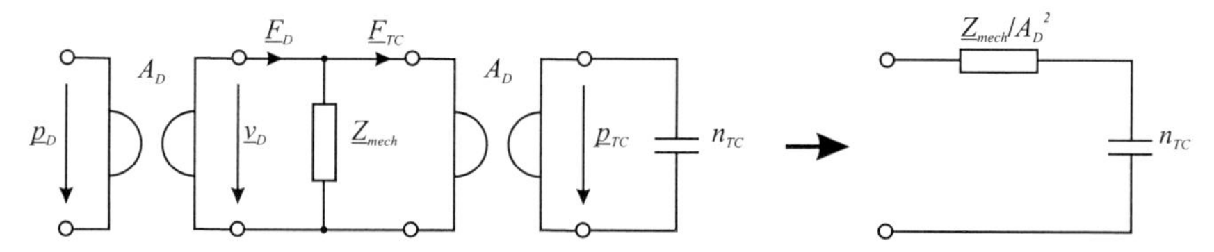

Fig. 10: Equivalent circuit demonstrating the influence of the tympanic cavity compliance n_{TC} on the eardrum impedance. The circuit on the left can be simplified. The resulting network shows that the stiffness of the air in the tympanic cavity is added to the mechanical eardrum impedance.

At the *stapes*, the model is terminated by a second order resonator which simulates the acoustical entrance impedance of the inner ear. The parameters were adapted according to Hudde and Engel, 1998abc. The translational degrees of freedom of the *stapes* are restricted in such a way that its footplate can only move in normal direction, while it is able to rotate freely.

Fig. 11 displays the meshed middle ear model and the inner ear element. The equivalent elements simulating the joints and tendons are not shown. The average element edge length on the tympanic membrane was adjusted to 0.8 mm, which seems to be sufficient to reproduce the fundamental eardrum vibration. Due to the high stiffness of the bone material, the distance between the FE nodes in the ossicles is not critical. The material parameters of the middle ear model are listed in Appendix A.2. The implemented model is designed to terminate the ear canal realistically, not to precisely simulate eardrum vibrations or middle ear characteristics. For that purpose, special models are available (Funnell and Decraemer 1996; Wada et al., 2002; Qi et al., 2008; Tuck-Lee et al., 2008 etc.). However, the calculation results of the model that are analyzed later are in good accordance with measurement data.

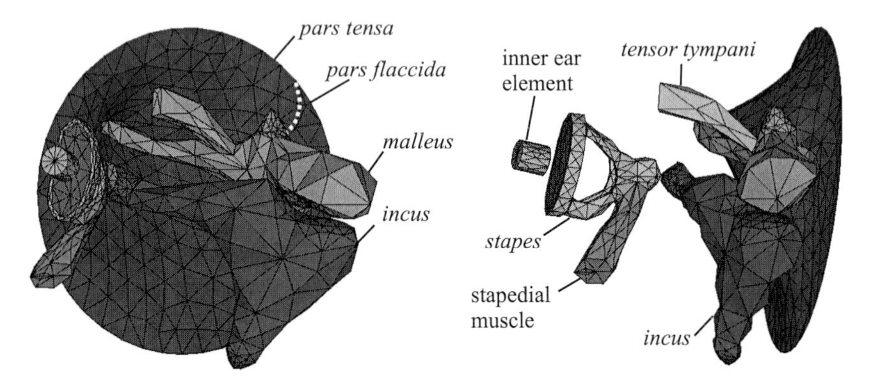

Fig. 11: Meshed model of the middle ear seen from two different viewing angles.

2.4 External ear sound fields

In this work, the structure of the sound fields is examined by plotting surfaces of equal pressure magnitude or phase angle ("isosurfaces") for given frequencies. In acoustical ducts with fundamental mode propagation, magnitude and phase isosurfaces are coincident and can be interpreted as "wavefronts" or ports of the duct. The isosurface plots are calculated by the FE software based on the simulated complex pressure phasors. The corresponding velocity field is depicted additionally, whenever it supports the discussion. According to Euler's equation, the spatial velocity is completely determined by the gradient of the sound pressure (Appendix A.1). Hence, it is a complex vector field. In general, the spatial vector components have separate phase angles. Instead of several frequency domain diagrams showing the vector components separately, the instantaneous velocity vectors at different time steps within the oscillation cycle are depicted.

2.4.1 Radiation impedance of the pinna

The pinna determines how external sound fields are coupled to the canal. In the network model approach described in section 2.2, its influence is included by the ideal pressure source $\underline{p}_{ecE}$ and the pinna radiation impedance $\underline{Z}_{ecE}$. The structure of the impedance function is determined by the pinna geometry, in particular for frequencies above approximately 4 kHz. Regarding the required measurements at the ear canal entrance, the relation between the pinna geometry and the radiation impedance was investigated. The ear canal was replaced with a straight tube by extruding the coupling area between the pinna and the ear canal along its normal. The posterior end of the tube was excited by a volume velocity source. The local acoustical impedance was determined from the resulting pressure according to the method described in section 2.5. The appropriate impedance transformation from the source to the entrance area of the ear canal was then determined according to a method proposed in Hudde and Engel, 1998b.

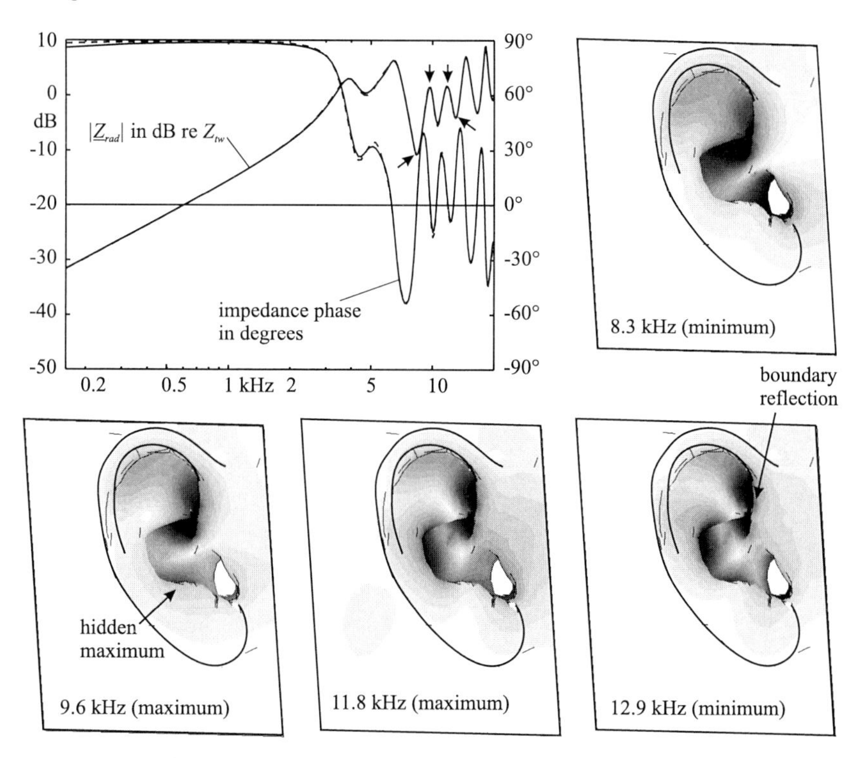

Fig. 12: Radiation impedance of the pinna (top left panel; normal pinna box: solid line, extended pinna box: dashed line) and sound pressure distribution on the boundary areas of the pinna (the corresponding frequencies are indicated by arrows in the impedance plot). The gray intensity corresponds to the pressure magnitude (black: pressure maximum, white: pressure minimum)

Fig. 12 (top left panel, solid lines) shows the resulting impedance. Up to 2 kHz, the impedance grows with 20 dB/decade and the phase angle is approximately 90°. The corresponding acoustical mass effect is caused by the fluid in the cavum conchae. At higher frequencies, pinna resonances are visible. The four pinna plots in Fig. 12 show the surface sound pressure distribution for the local impedance minima and maxima that are indicated by the arrows in the impedance plot. Pressure magnitude values are indicated by different gray values (black: pressure maximum, white: pressure minimum).

The figures show that the resonances originate from strong reflections at the boundary indicated in the plot for 12.9 kHz (*crus helicis*). At 8.3 kHz (impedance minimum), two distinct maxima are visible (one at the *crus helicis*, the other in the deep *cavum* at the transition to the canal). At 9.6 kHz (impedance maximum), only one maximum seems to exist, a second, however, is located behind the ridge indicated in the plot (*antitragus*) and is thus not visible. At 11.8 kHz and 12.9 kHz, three and four maxima are visible. The results suggest that the structure that bounds the *cavum* (*antihelix*) acts as acoustical duct with a rigid termination at the *crus helicis.*

In addition, it was tested whether the size of the pinna box influences the radiation impedance results. The small box with dimensions of 80 x 70 x 50 mm was extended to edge lengths of 130 x 100 x 70 mm. Thus, the volume was increased by a factor of 3.25. Fig. 12 (top left panel, dashed lines) shows the resulting impedance. Obviously, enlarging the volume of the pinna box has no significant impact on the radiation impedance. In the following, only the small pinna box is used.

2.4.2 The sound field near the tympanic membrane

To evaluate the eardrum simulation, its vibrational pattern and the resulting eardrum impedance are compared with measurements (Fig. 13). In the left panel, the displacement magnitude for an excitation frequency of 960 Hz and a local sound pressure of 1 Pa is displayed. It shows significant accordance with the holographic measurements documented by Tonndorf and Khanna, 1972. Here, a harmonic sound pressure of 1 Pa and a frequency of 996 Hz excited a displacement of $6.7 \cdot 10^{-8}$ m at the *umbo*. In the FE model, a displacement of $4.2 \cdot 10^{-8}$ m is calculated at 960 Hz for the same pressure. Regarding the simplifications, the difference between the measurement and the model prediction is very small.

A good congruence is given as well with the data of Gyo et al., 1987, and Kringlebotn and Gundersen, 1985, who present measurements of middle ear data. In the references, the transfer function between the pressure at the *umbo* and the *stapes* displacement has been determined as function of frequency (the displacement magnitudes of *stapes* and *umbo* are only marginally different). The center panel of Fig. 13 shows a comparison with the model presented in the work at hand. The transfer function documented by Kringlebotn and Gundersen provides a significant peak at 900 Hz, and both measurements show a decrease to higher frequencies. These features are represented in the model data as well. In addition, the acoustical impedance of the modeled tympanic membrane was compared with the network model proposed by Hudde and Engel, 1998abc (right panel of Fig. 13). The parameters of the Hudde and Engel model were adjusted to measurements carried out in human cadaver ears. The results show that the eardrum is rather stiff, the impedance is always approximately 10 dB larger than the tube wave impedance of a duct with similar diameter. Except for a peak near 2 kHz that is not reproduced by the FE model, the difference between the results is very small. The origin of the difference is unclear. It can be supposed that the damping of the FE model has more effect than in the network model. In conclusion, the eardrum model is suited well for the purposes of the present investigation.

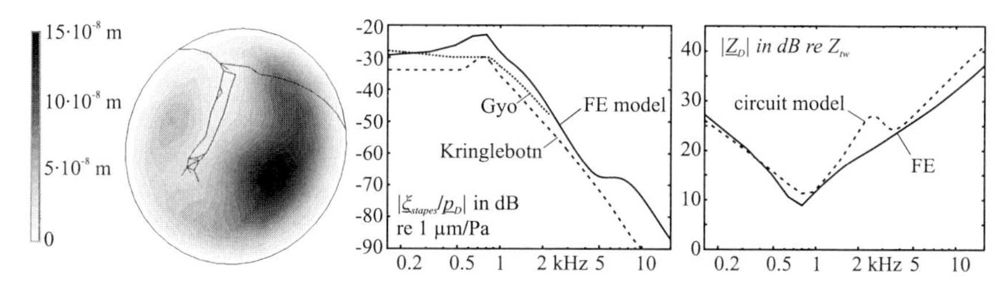

Fig. 13: Vibrational pattern of the tympanic membrane at f=960 Hz and $|p_T|=1$ Pa (left panel), transfer function between the pressure at the umbo and the stapes displacement (center panel) and acoustical impedance of the model eardrum in comparison to the circuit model by Hudde and Engel, 1998abc (right panel), normalized by the tube wave impedance Z_{tw} of a straight duct with the same cross-sectional area as the ear canal.

In most ears, the tympanic membrane is inclined significantly against the middle axis of the canal, thus, it smoothly merges with the ear canal wall at its top. It can be assumed that waves are guided parallel to the eardrum due to its stiffness. This was evaluated by isosurface plots of the sound field in the posterior ear canal section. As expected, the field structure near the eardrum is independent from the source. When the direction of incident sound waves is varied (frontal, dorsal and lateral source, see subsection 2.3.2), no visible differences can be found. Thus, only one of the three cases needs to be examined.

In Fig. 14, pressure magnitude values are indicated by different gray values (black: pressure maximum, white: pressure minimum). For each frequency, the pressure range occurring in the depicted region is divided into 20 equidistant steps. As a matter of course, various frequencies yield different magnitude minima and maxima. Hence, the pressure difference between two adjacent isosurfaces varies between the plots. Additionally, it depends on the value of the excitation volume velocity. On the other hand, the applied method makes the shape and orientation of local isosurfaces better comparable for different frequencies. For a notion of the depicted pressure gradients, the dynamic range $D=20{\cdot}\log_{10}|p_{max}/p_{min}|$ of the pressure field is given beneath the respective panel.

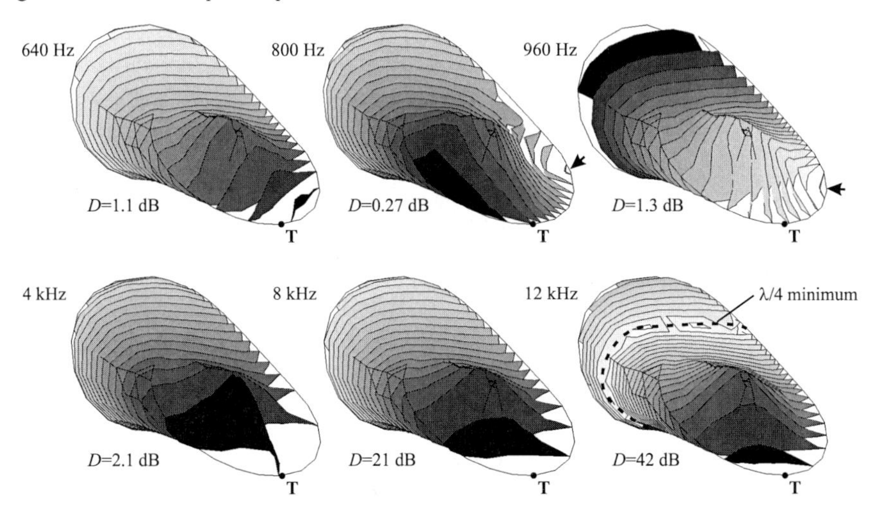

Fig. 14: Isosurfaces near the tympanic membrane for frontal sound incidence (other incidence direction result in almost identical structures). The excitation frequency is given above the respective plot; the value $D=20{\cdot}\log_{10}|p_{max}/p_{min}|$ denotes the dynamic sound field range.

For the first three frequencies, the pressure magnitude differences in the eardrum coupling region are very small. At low frequencies, the wavelength is large against the canal. It can be modeled as lumped element in good approximation. Between 4 kHz and 12 kHz, pressure magnitude isosurfaces are aligned orthogonal to the eardrum surface. As expected, the waves are guided parallel to the tympanic membrane into the tympanomeatal angle between the eardrum and the canal wall and consequently towards the lower edge of the eardrum (point **T**). Here, a pressure reflection without sign change is obtained. This supports the basic assumptions concerning the eardrum coupling region of the one-dimensional model described in section 2.2. At 12 kHz, the quarter wavelength minimum is located approximately 10 mm anterior to the point **T**. In a cylindrical duct, the corresponding minimum for the same frequency would be found at 7.2 mm. The difference originates from the conical form of the eardrum coupling region.

As it is rather stiff for higher frequencies, the eardrum vibrations do not contribute significantly to the sound field in its direct vicinity. This becomes more obvious, when the isosurface patterns of Fig. 14 are compared with isosurfaces that arise when the tympanic membrane is modeled absolutely rigid (see Fig. 15). Here, incident waves are reflected at a rather constant location in the innermost part of the tympanomeatal corner. This supports the concept of the point **T** as termination and reflection port. In contrast to Fig. 14, the isosurface structure remains almost constant for the selected frequencies. Distinct differences between the rigid and the normal eardrum can be observed in the frequency range between 600 Hz and 4 kHz. The arrows near the tympanomeatal corner that can be found in the panels of Fig. 14 indicate local pressure minima. Astoundingly, a minimum occurs in the tympanomeatal corner for 960 Hz. In this particular frequency range, the eardrum vibration essentially affects the local sound field, because the main resonance of the middle ear and several resonances of the eardrum occur. Comparing both cases, it becomes further visible that the effect of the eardrum vibration becomes less influence for the frequencies 640 Hz and 4 kHz.

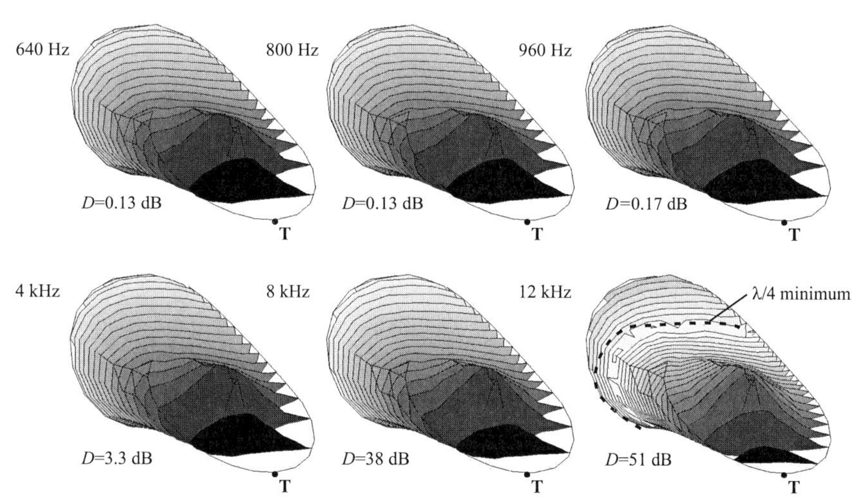

Fig. 15: Isosurfaces corresponding to Fig. 14 for the case of a rigid tympanic membrane

At 12 kHz, both a magnitude minimum and maximum are located in the region next to the rigid tympanic membrane. Hence, the dynamic range D is identical with the standing wave ratio of the system. When a straight wave duct with planar wave propagation is terminated rigidly, the standing wave ratio becomes infinite, as the incident and reflected wave perfectly cancel each other in the minimum. Due to numerical errors, an FE model of this configuration obtains a finite standing wave ratio. Using an example simulation of a cylindrical duct, a standing wave ratio of D=104 dB could be achieved. However, the dynamic range of the field in the rigidly terminated ear canal model has a significantly smaller value (D=51 dB) which indicates that the minimal pressure is different from zero. It can be assumed that the interference of incident and reflected wave in the ear canal is not totally destructive, because the wave fronts have slightly different shape. This issue will be addressed further in subsection 2.4.6.

The sound field arising in a natural ear canal consists of a homogeneous part that is represented by the isosurface structure that would arise at a rigid eardrum and a superimposed disturbance caused by the eardrum vibration. As result, irregular three-dimensional shapes occur and vary with frequency. As one-dimensional modeling approach, a frequency-dependent middle axis could be implemented. For 640 Hz, 960 Hz and 4 kHz, the axis could be determined from the isosurfaces. For 800 Hz, both a minimum and a maximum occur on the eardrum rim. This case can hardly be modeled generally.

Below 1 kHz, however, the wavelength is large enough to regard the ear canal as lumped element with an approximately constant spatial sound pressure distribution. For frequency ranges in which the wavelength of sound becomes comparable to the canal dimensions and thus wave effects have to be considered, the sound field disturbance of the tympanic membrane can be neglected, as was shown in the figures above. In Fig. 16, ratios of various pressure signals are depicted to quantify the effect of eardrum vibrations. It becomes obvious that the maximum deviation between the pressure signal at the point **T** and the actual global maximum in the eardrum coupling region increases only about 1 dB at the middle ear and eardrum resonance. At higher frequencies, the difference of the pressure signals continuously decreases in spite of the larger field dynamic in the eardrum coupling region, because the maximum location approaches the point **T**.

In the right panel of Fig. 16, differences between pressure spectra at rigid and normal eardrums are documented. The dashed line belongs to the point **T**, the dash-dotted line to the isosurface at **X** (approximate entrance of the eardrum coupling region). The variations of up to 10 dB show that the pressure field is essentially influenced, when strong eardrum vibrations are prominent (up to approximately 3 kHz). For higher frequencies, the pressure in the tympanomeatal corner is affected only marginally by the eardrum (variations of 3 dB). The large differences near 8 kHz that occur in the curve assigned to the surface **X** are caused by a pressure minimum which is slightly displaced by the change in the termination impedance of the ear canal.

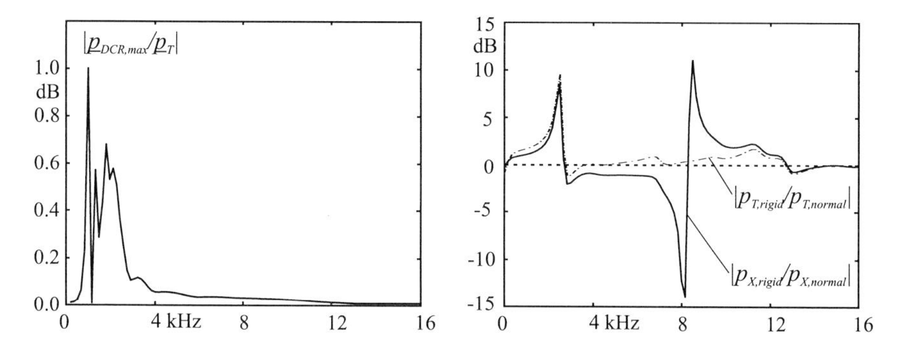

*Fig. 16: Pressure level differences in the eardrum coupling region. Left: level difference between the pressure magnitude at the termination point **T** and the absolute pressure maximum occurring in the eardrum coupling region for the respective frequency, calculated for a normal eardrum. Right: Differences between normal and rigid eardrums at different locations in the canal.*

In conclusion, the effect of eardrum vibrations on the sound field is only significant for low frequencies, where the eardrum coupling region can be approximated as lumped element. It is thus reasonable to specify the pressure $\underline{p}_T$ arising at the point **T** as eardrum signal in the sense of this thesis. In the following, it is investigated, how the transfer functions between $\underline{p}_T$ and the inner ear signals ear depend on the local anatomy, in particular the eardrum inclination angle.

2.4.3 Transfer functions to the middle and inner ear

For the reference of psychoacoustical measurements to the eardrum signal $\underline{p}_T$, it is required that transfer functions between $\underline{p}_T$ and signals in the middle and inner ear depend on the individual eardrum geometry only to a small extent.

In the following, transfer functions between $\underline{p}_T$ and several quantities calculated from the pressure distribution over the tympanic membrane are analyzed. As the tympanic membrane is clamped at its edge, it does not vibrate as a whole, rather as an elastic membrane. The transmission of sound into the middle ear is obtained by the coupling area of the *malleus*. Due to the membrane stiffness of the tympanic membrane, *malleus* movements can be excited by forces acting upon any point of the eardrum surface, except for the rim. Thus, three pressure signals are calculated: average pressure $\underline{p}_{A,D}$ over the total eardrum surface, average pressure $\underline{p}_{A,umbo}$ over the central part of the eardrum near the *umbo* and point pressure $\underline{p}_{umbo}$ at the *umbo*.

To specify the relation to the inner ear signals, the pressure in the vestibule ($\underline{p}_V$) is evaluated. The latter can be computed from the *stapes* displacement ξ_{stapes}. Using the acoustical impedance of the vestibule $\underline{Z}_V$ (from Hudde and Engel, 1998abc) and the surface area of the *stapes* plate A_{stapes}, it holds $\underline{p}_V=\underline{q}_V\underline{Z}_V=\underline{v}_{stapes}\underline{Z}_V A_{stapes}=j\omega\underline{\xi}_{stapes}\underline{Z}_V A_{stapes}$.

For this investigation, the model was simplified to reduce the computational expense. As shown above, the field structure in the eardrum coupling region does not depend on the source. Thus, pinna box and ear canal could be replaced with a straight duct and a volume velocity source. Nine models with different angles of eardrum inclination against the approximate middle axis in the eardrum coupling region were implemented (20° to 90° in steps of 10°, in addition 45°) The simulation frequencies ranged up to 20 kHz in equidistant steps of 160 Hz.

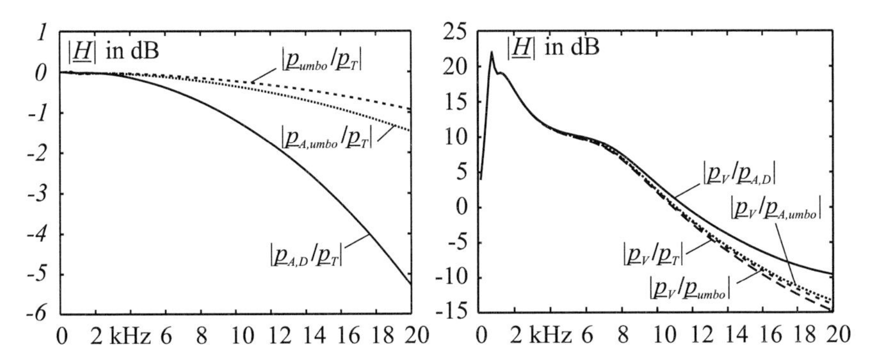

Fig. 17: left: transfer functions between $\underline{p}_{A,D}$, $\underline{p}_{A,umbo}$ and $\underline{p}_{umbo}$ and $\underline{p}_T$ (for 45° eardrum inclination), right: transfer functions between the pressure signals at the eardrum and the vestibule pressure $\underline{p}_V$.

In Fig. 17 (left panel), transfer functions between $\underline{p}_{A,D}$, $\underline{p}_{A,umbo}$ and $\underline{p}_{umbo}$ and the pressure $\underline{p}_T$ are shown (only for the case of 45° eardrum inclination). For frequencies below 4 kHz, the pressure over the whole eardrum surface can be regarded as constant. Obviously, $\underline{p}_T$ differs only marginally from the two signals related to the *umbo* (max. -1.5 dB at 20 kHz), whereas the integration of pressure over the total eardrum surface yields a larger error (max. -5.5 dB at 20 kHz). This can be explained by the large difference pressure that arises over the tympanic membrane for high frequencies due to the small wavelength (see Fig. 14, panel for 12 kHz, e.g.). The right panel of Fig. 17 shows the transfer functions between the signals $\underline{p}_{A,D}$, $\underline{p}_{A,umbo}$

and $\underline{p}_{umbo}$ and the vestibule pressure $\underline{p}_V$. As the FE model represents a linear system, the curves have the same attributes as in the left panel: the transfer characteristics are almost identical up to 4 to 6 kHz, for higher frequencies differences become visible. The largest deviation is obtained, when the transfer function is related to the pressure averaged over the whole eardrum area.

In the following, the influence of the eardrum geometry on the transfer functions described above is evaluated. The tympanic membrane with an inclination of 45° against the middle axis is used as reference case. In the plots in Fig. 18, Fig. 19 and Fig. 20, the ratio $|\underline{H}_\alpha/\underline{H}_{45°}|$ of the transfer functions occurring in the reference case and for the variations (α=20° ... 90°) is displayed. Additionally, dotted lines at ±1 dB deviation are shown. In the bottom rows, the phase response of the transfer functions is plotted. The graphs in Fig. 18 and Fig. 19 suggest that the transfer functions between signals related to the eardrum surface and $\underline{p}_T$ are influenced by the eardrum inclination angle only to a small extent.

For the same reasons as stated above, the transfer characteristics are influenced more, when the pressure integral over the tympanic membrane is taken as reference. The results displayed in Fig. 20 show that the transfer function $\underline{H}=\underline{p}_V/\underline{p}_T$ is robust against geometry variations of the eardrum. Only for extremely steep eardrum geometries (80°-90°), the difference to the reference case rises above 1 dB.

In conclusion, the pressure occurring at the *umbo* optimally represents the input to the middle ear. The transfer function between the tip of the *umbo* or an area nearby and the inner ear is practically independent of the tympanic membrane geometry. Thus, it would be suited optimal as a reference signal for audiological measurements; however, it cannot be measured precisely for reasons already discussed in the introduction. The investigation shows that the transfer function between the tympanomeatal corner and the vestibule is similarly robust against eardrum geometry changes. In addition, the differences between the signals $\underline{p}_T$ and $\underline{p}_{umbo}$ can be neglected. This confirms that $\underline{p}_T$ can be interpreted as input to the middle ear.

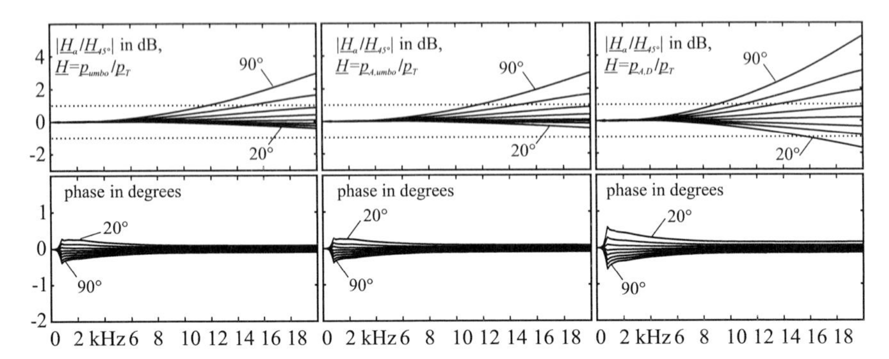

Fig. 18: Influence of the eardrum inclination angle (α=20° ... 90°) on the eardrum transfer functions $\underline{H}=\underline{p}_{umbo}/\underline{p}_T$, $\underline{H}=\underline{p}_{A,umbo}/\underline{p}_T$ and $\underline{H}=\underline{p}_{A,D}/\underline{p}_T$ with reference to the 45° case.

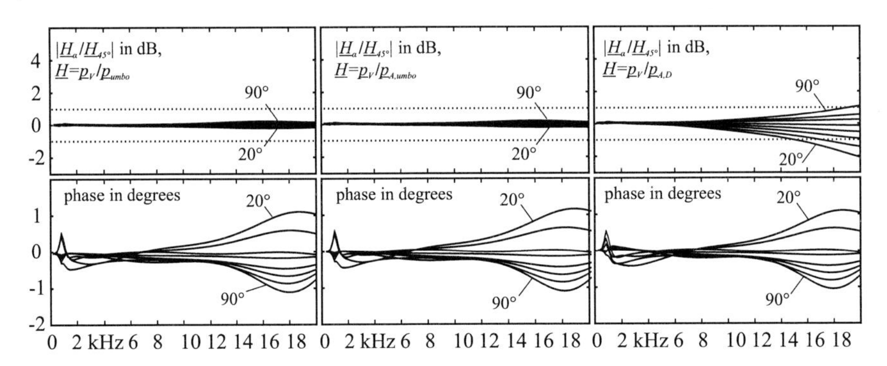

Fig. 19: Influence of the inclination angle of the eardrum (α=20° ... 90°) on the transfer functions $\underline{H}=\underline{p}_V/\underline{p}_{umbo}$, $\underline{H}=\underline{p}_V/\underline{p}_{A,umbo}$ and $\underline{H}=\underline{p}_V/\underline{p}_{A,D}$ with reference to the 45° case.

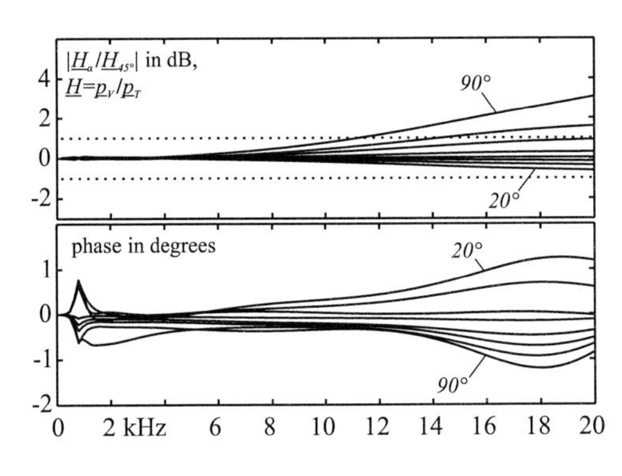

Fig. 20: Influence of the inclination angle of the eardrum (α=20° ... 90°) on the transfer function $\underline{H}=\underline{p}_V/\underline{p}_T$ between eardrum and vestibule pressure with reference to the 45° case.

2.4.4 Sound fields within and outside the ear canal

The ear canal sound field can be excited by very different sources. For example, remote sources in anechoic conditions excite approximately unidirectional plane waves at the listener's position. In reverberant conditions or in the near field of a source, complex spatial fields arise in front of the pinna. The external sound field is a superposition of the source field (sound waves that would be present if the listener was absent), reflection at the external ear (including torso, head and pinna), and diffraction effects of the head. Sources can be positioned very close to the pinna (earphones) or even within the ear canal (hearing aids). For the estimation of $\underline{p}_T$ from measurements in the ear canal, it must be known up to which point sound sources influence the field structure in the ear canal (as a matter of course, the source signal determines the pressure magnitude at each point; here, the influence on the field geometry is considered). At a particular point in the ear canal, the shape of isosurfaces can be assumed to be independent from the three-dimensional field near the pinna. In the following, the transition region between the external and internal sound fields is examined to study the interaction of the two field types.

In this investigation, the pinna box (subsection 2.3.2) is applied. As one would expect, the isosurface structure in front of the pinna is essentially determined by the excitation direction (dorsal, lateral or frontal). In Fig. 21, the respective isosurface patterns arising at 320 Hz are depicted. The shape of isosurfaces in the pinna box plotted in the top row shows that the model simulates different sound incidence directions acceptably at low frequencies. For dorsal and frontal excitation, grazing waves occur, while impinging wave fronts are obtained using the lateral source. For higher frequencies, the structures become more complicated.

In the middle row, a horizontal cross-sectional area of the pinna box is shown. The section surface has the same vertical position as the middle axis at the ear canal entrance. Thus, the external and internal field and in particular the transition region become visible. Pressure magnitudes in the cross-sectional area are given as gray tone intensities. Again, the dynamic range D of the displayed field is given next to the respective plot.

In the bottom row, isosurfaces in the ear canal are shown. For each excitation direction, the arrow above the ear canal shows an isosurface that is common for the three cases. As expected, the field in the central ear canal region is independent from the source, at least for the shown frequency. To the left of the arrow, the isosurface shapes of the dorsal and frontal incidence model are comparable, whereas the lateral incidence case shows distinct differences. The core region is independent of the source by definition, thus, the arrows indicate its entrance for the given frequency. Anyway, at 320 Hz, the pressure magnitude variation in the simulation region are extremely small, thus, the whole system can be accurately modeled by lumped element approximations.

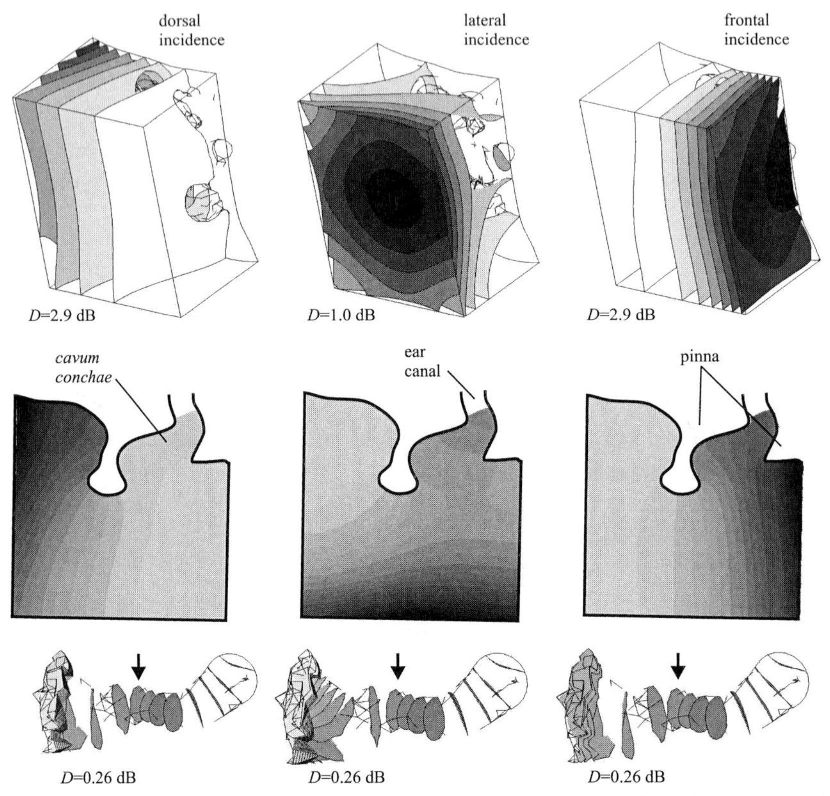

Fig. 21: External ear sound fields for dorsal, lateral and frontal sound incidence at 320 Hz. The sound pressure magnitude is indicated by the intensity of gray tones (darker values correspond with larger magnitudes). Top: isosurfaces in the pinna box. Center: pressure contours in a horizontal cross-sectional area of the pinna box. Bottom: pressure isosurfaces in the ear canal. The arrows show the first isosurface that can be regarded as independent from the source.

However, at higher frequencies, pronounced wave effects can be observed. Due to duct effects in the ear canal, pressure and volume velocity extrema appear near the entrance for particular frequencies. It is important to examine the field structure for these cases, as it affects the coupling of the external field and the canal essentially. In ducts, pressure minima generally correspond with velocity maxima and impedance minima (and vice versa). The first two pressure extrema at the ear canal entrance can be found at 4480 Hz (minimum) and 9120 Hz (maximum). The slice contours or the pressure distribution and the ear canal isosurface plots for 4480 Hz show that the sound field geometry does not depend on the pressure distribution in the pinna box (Fig. 22, top). This holds even in the *cavum conchae*, where the sound field appears to be similar for the three cases, although the excitation field is different. The corresponding velocity maximum (Fig. 23, top) is indicated by the long vectors at the canal entrance. The velocity directions are almost identical for the three sources and constant over time. After a half period $T/2$, merely the sign is changed. The large velocity strongly couples the fields in the concha and the ear canal. The resulting isosurface shapes are nearly independent of source variations, even at a short distance outside the ear canal in the *cavum conchae*.

At 9120 Hz, the frontal and dorsal sources excite a similar sound field, whereas the lateral source builds up a very different and fairly irregular field structure. Dome-shaped structures occur which can hardly be aligned with the middle axis of the canal. As indicated by the arrows in Fig. 22, the entrance of the core region is significantly shifted to a posterior position. In the *cavum*, the pressure becomes maximal. Consequently, a velocity minimum is present at the entrance (Fig. 23, four bottom panels).

The rigid termination condition at the end of the ear canal is transformed to the location of the velocity minimum, which thus represents a surface with high acoustical impedance. As the pressure is not subject to any boundary conditions here, the sound fields are coupled only weakly. The source influence in pressure minima reaches up to more posterior positions in the field, which has essential consequences for measurements at the ear canal entrance (this issue will be examined in section 2.5).

This finding is supported by the instantaneous velocity vectors of the canal field that are visualized in the four bottom panels of Fig. 23 for four arbitrary time instants (t_A to t_D). The upper row shows data for the fontal excitation, whereas the lower row depicts the lateral source case (the dorsal case yields similar results). To the left of the velocity minimum, four significantly different vector fields are visible for the frontal and lateral source. Here, the velocity arrows in the concha (to the left of the minimum) continuosly change their orientation during one cycle of oscillation. The vector fields in the canal show only marginal variations, which is an attribute of the core region.

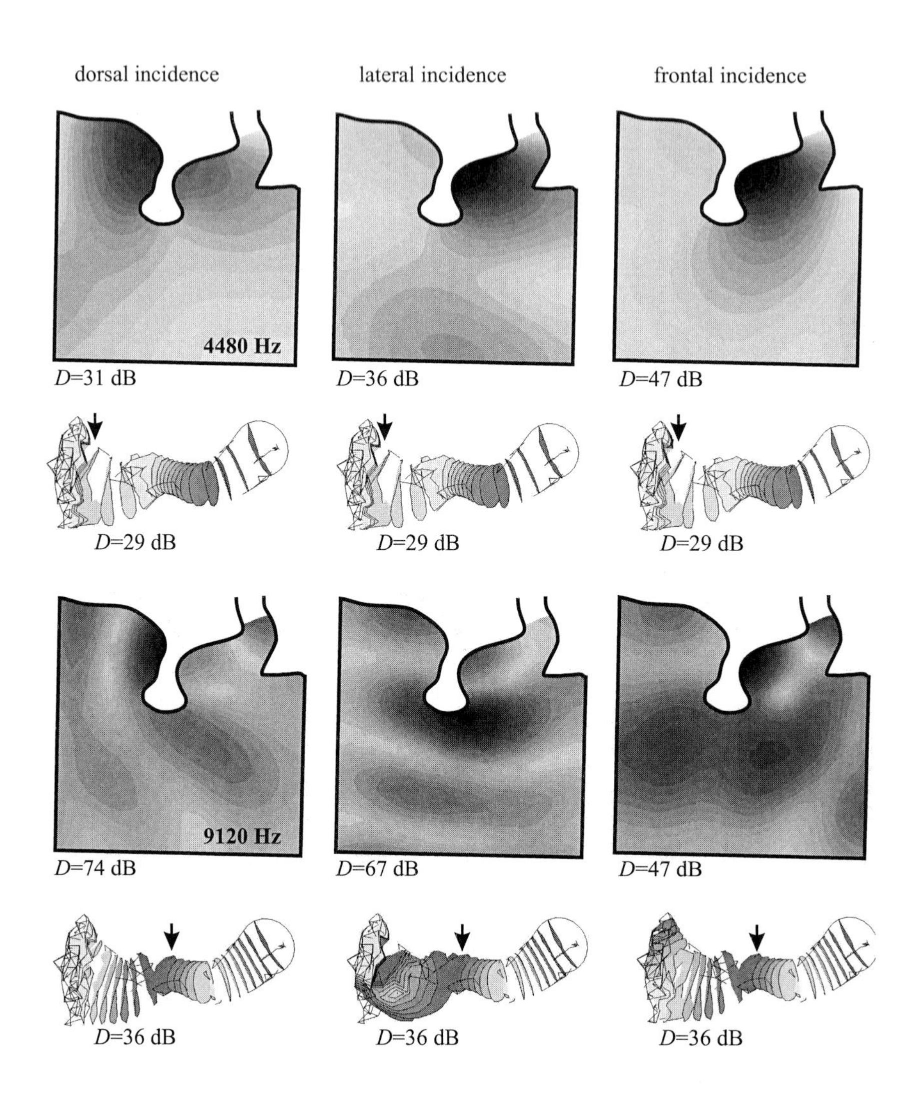

Fig. 22: External ear sound fields for dorsal, lateral and frontal sound incidence at 4480 Hz (top) and 9120 Hz (bottom) depicted as pressure magnitude contours. In the ear canal isosurface diagrams below the cross-sectional plots, the first isosurfaces that can be regarded as independent from the source are indicated by arrows.

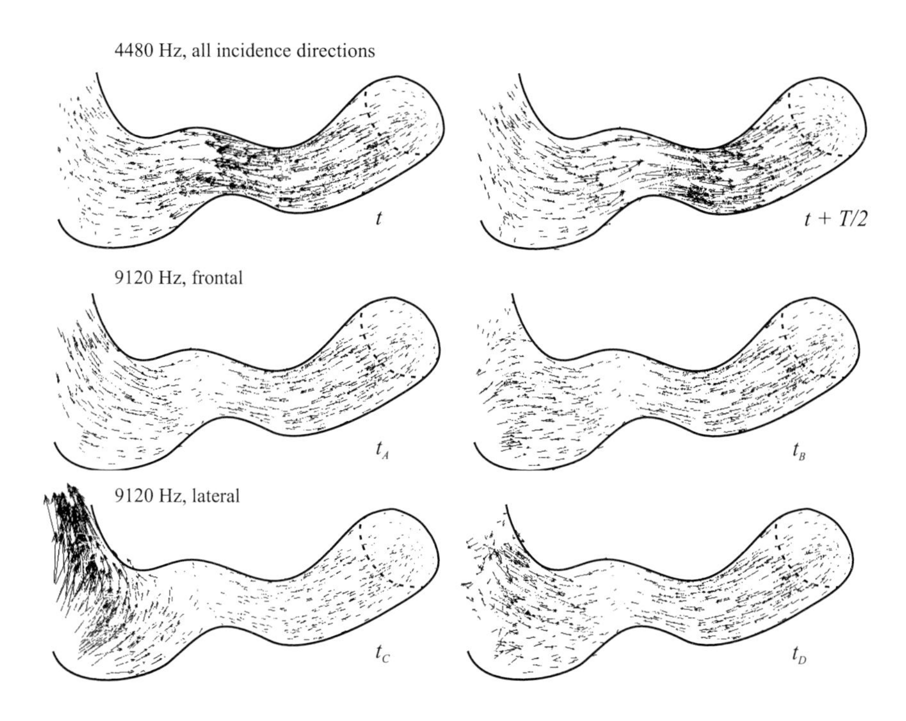

Fig. 23: Instantaneous velocities in the ear canal. Top: The orientation of the velocities at 4480 Hz is almost identical for the three directions of incidence and constant with time. Other four panels below: At 9120 Hz the orientation of the velocity vectors in the concha depends on the sound incidence direction. In addition, it varies during a cycle of the oscillation (time instants t_A to t_D).

In pressure minima that occur at higher frequencies, a coupling as strong as for 4480 Hz cannot be found.With decreasing wavelength, the distance between two adjacent minima and maxima decreases as well. Several minima and maxima can be found in the *concha*. Hence, a robust discrimination of strong or weak coupling becomes ambiguous. The complexity of the sound field increases. Examples for 14240 Hz are visualized in Fig. 24. Generally, the entrance position of the core region is fairly constant for higher frequencies. In straight sections of the canal, the isosurfaces seem to be regular (bottom row), whereas complicated shapes arise in the concha and in the curves. This kind of isosurfaces is addressed in the next section.

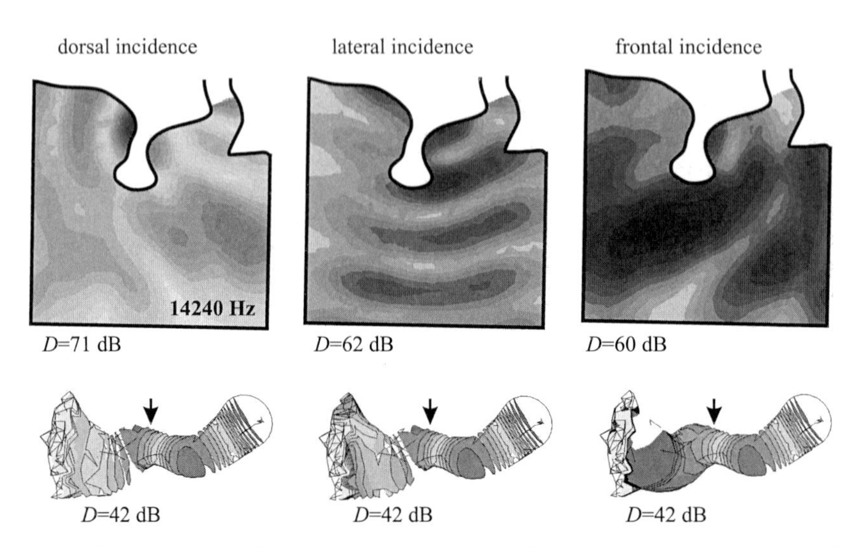

Fig. 24: External ear sound fields for dorsal, lateral and frontal sound incidence at 14240 Hz.

2.4.5 One-sided isosurfaces

In some of the sound fields depicted in the previous section, a special type of isosurfaces can be observed. At concave regions of the canal walls (such as the curves), the surfaces form a set of domes that span over only one side of the ear canal. Hence, they are called "one-sided isosurfaces" in the following. Such structures are present in Fig. 22, lower plots, and Fig. 24. The other type of isosurfaces which covers the entire cross-section of the ear canal is called "regular".

One-sided isosurfaces systematically arise in curved ducts. They occur at the duct wall, whenever the middle axis is not totally straight as in axisymmetric ducts. The sound pressure is slightly higher on the concave side of the curve. Consequently, a local pressure maximum arises on the concave side, whereas a local pressure minimum will be positioned on the convex side. One-sided isosurfaces are arranged in layers around the points of minimal or maximal sound pressure magnitude, which are degenerated "isosurfaces".

To study isolated one-sided isosurfaces, a simplified FE model of a duct was designed. The model consists of two circular cylindrical tubes which are connected by a toroidal bend constructed using splines. The duct is excited by a volume velocity source with internal impedance at one end. The other end is terminated with the double tube wave impedance of the cylindrical duct. The simulation frequency is adjusted in such a way that a pressure maximum occurs in the bending. A pressure minimum at the same position is simply obtained by changing the termination impedance to its inverse. The lateral dimensions of the duct are comparable with ear canals (radius 4 mm), its total length was set to approximately 60 mm. The resulting magnitude and phase isosurfaces are depicted in Fig. 25.

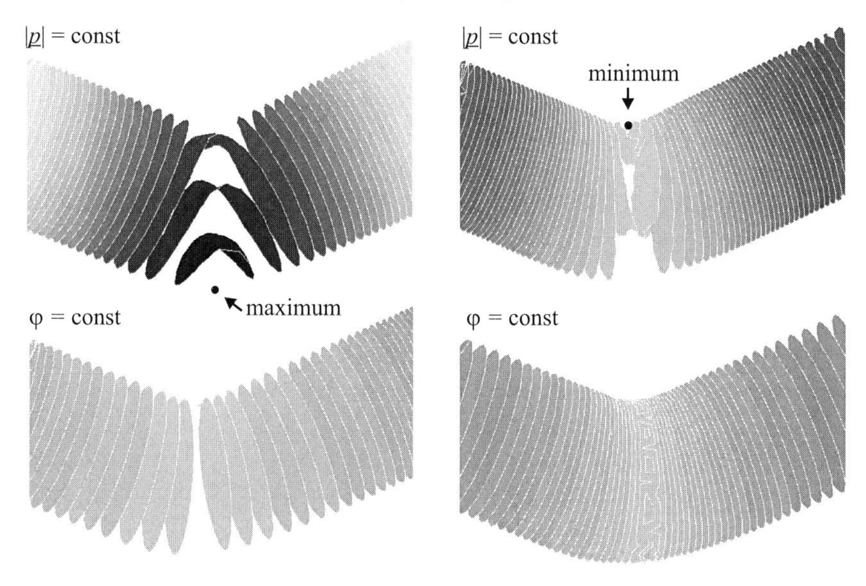

Fig. 25: Isosurfaces for a pressure maximum (left column) and minimum (right column) occurring in a curved duct. Top row: magnitude isosurfaces, bottom row: phase isosurfaces. As the difference of magnitude or phase between two adjacent isosurfaces is constant, the higher density of surfaces near the minimum indicates larger gradients compared to the maximum.

As mentioned above, the maximum arises in the concave bending, whereas the minimum appears on the convex side (left and right panels in top row, respectively). In the attached straight ducts, a field of regular isosurfaces is visible. The pressure maximum appears to be broader than the minimum, which is a general attribute of pressure maxima in ducts in which the incident waves are strongly reflected at the end.

Generally, one-sided isosurfaces may be found at all frequencies, but their shape and position are not constant. They are generated by local evanescent higher-order modes, thus, they cannot be decomposed into fundamental mode waves. To investigate how one-sided isosurfaces influence fundamental mode fields, it is necessary to take the velocity field around one-sided magnitude isosurfaces and the corresponding phase isosurfaces into account.

According to Euler's equation (Appendix A.1), the velocity vector field can be calculated by the field gradient of the pressure distribution. Assuming a monofrequent pressure signal $\underline{p}(\mathbf{r}) = |\underline{p}(\mathbf{r})| \mathrm{e}^{\mathrm{j}\varphi_p(\mathbf{r})}$, the velocity vector at the position $\mathbf{r}$ results to

$$\underline{\mathbf{v}}(\mathbf{r}) = -\frac{\mathrm{grad}\{\underline{p}(\mathbf{r})\}}{\mathrm{j}\omega\rho} = \frac{\mathrm{e}^{\mathrm{j}\varphi_p(\mathbf{r})}}{\omega\rho}\left(j \cdot \mathrm{grad}\{|\underline{p}(\mathbf{r})|\} - |\underline{p}(\mathbf{r})| \cdot \mathrm{grad}\{\varphi_p(\mathbf{r})\}\right) \tag{2.4.1}$$

This equation shows that the direction of velocity is not exclusively determined by the pressure magnitude gradient which is oriented normal to the pressure magnitude isosurfaces ($|\underline{p}|$=*const*). The vectors also depend on the phase gradient which is orthogonal to the phase isosurfaces, weighted with the local pressure magnitude. The phase isosurfaces in the model duct (bottom row of Fig. 25) are nearly parallel and indicate a continuous and monotonous phase development and consequently the wave propagation direction. On regular magnitude (or phase) isosurfaces, all point pressures oscillate in phase; the velocities at all points have constant orientation and are normal to the surface. In this case, the wave propagation direction is identical with the orientation of the velocity vectors. It is thus reasonable to define the local section of the middle axis along the velocity direction given in the isosurface area centroid. Equation (2.4.1) also obtains that no velocity components normal to rigid walls can arise, because both gradient terms are zero in the direction perpendicular to the walls. Consequently, both regular and one-sided isosurfaces are always orthogonal to rigid walls.

The orientation of velocity vectors is generally not constant during one cycle of oscillation which becomes obvious, when Eq. (2.4.1) is formulated in time domain:

$$\begin{aligned} &\underline{\mathbf{v}}(\mathbf{r},t) = \mathrm{Re}\{\underline{\mathbf{v}}(\mathbf{r})\mathrm{e}^{\mathrm{j}\omega t}\} = \\ &-\frac{1}{\omega\rho}\left[\mathrm{grad}\{|\underline{p}(\mathbf{r})|\}\sin\left(\omega t + \varphi_p(\mathbf{r})\right) + |\underline{p}(\mathbf{r})|\,\mathrm{grad}\{\varphi_p(\mathbf{r})\}\cos\left(\omega t + \varphi_p(\mathbf{r})\right)\right] \end{aligned} \tag{2.4.2}$$

As both the magnitude and phase gradient term are multiplied with orthogonal sinusoidal functions, the instantaneous direction of the vectors oscillates between the directions of the two gradients (actually, this finding gave rise to the use of time domain visualization of vector plots in this thesis). Thus, exclusively for coinciding isosurfaces of magnitude and phase, the direction of the velocities is time-invariant and normal to the isosurfaces. In this case, the orientation of instantaneous velocity vectors is constant and only the sign is reversed. According to the phase isosurfaces in the model duct (bottom row of Fig. 25), the velocities in pressure maxima are oriented parallel to the middle axis. Only when the pressure magnitude is very small, the magnitude gradient exclusively determines the direction of velocity according to Eq. (2.4.2). In such time instants, the velocity vectors rapidly turn into the opposite direction. This process can be observed in the four panels of Fig. 26. The direction is reversed at the time instant t_0. This event is repeated after each half cycle $T/2$ under inverse conditions. At t_0, the velocity in the pressure maximum is directed normal to the walls, except immediately at the walls where the normal velocity is zero. Hence, the velocity direction in pressure maxima in essentially determined by the pressure phase angle.

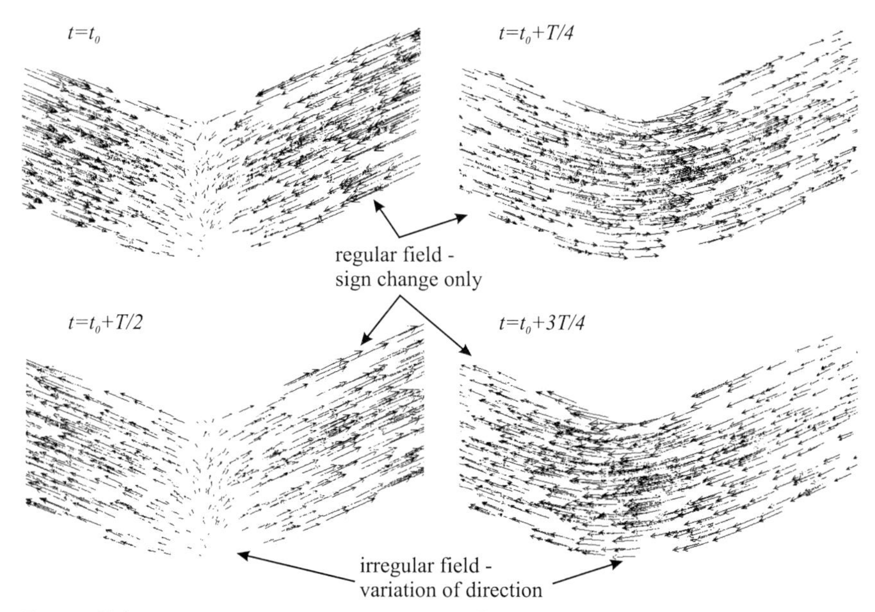

Fig. 26: Velocity vectors near a pressure magnitude maximum in a curved duct for four time instants in a cycle of period T. At $t=t_0+kT/2$*, k being integer, the otherwise nearly constant direction of velocity vectors is reversed.*

The characteristics of one-sided isosurfaces originated by a pressure magnitude maximum are summarized schematically in Fig. 27. Sufficiently far from the maximum the isosurfaces of magnitude and phase coincide. Approaching the extremal value, the magnitude and phase gradients differ in direction. Hence, the velocity vectors are not normal to the magnitude isosurfaces. Near a pressure maximum, the direction of the velocity is mainly determined by the phase gradient. The direction of propagation is thus obtained even when the pressure magnitude vanishes, as it is identical to the phase gradient.

As pressure minima are generally narrow compared to maxima, the pressure values increase over a shorter distance. Thus, the magnitude gradients are significantly larger. This is represented by the smaller distance of isosurfaces in Fig. 25. As the isosurfaces are spaced in equidistant magnitude or phase steps, the respective gradient is reciprocal to the spatial isosurface distance. An inspection of the velocity field yields that in contrast to the broad maximum, a significant component of the velocity vectors points towards the minimum. Consequently, it is oriented almost orthogonal to the direction of wave propagation which seems to contradict a definite propagation direction. However, Fig. 25 shows that the isosurfaces approach regular surfaces already very close to the minimum. Further, the corresponding phase isosurfaces are aligned closer as for the pressure maximum. This results in a larger phase gradient which constitutes the orientation of velocity vectors.

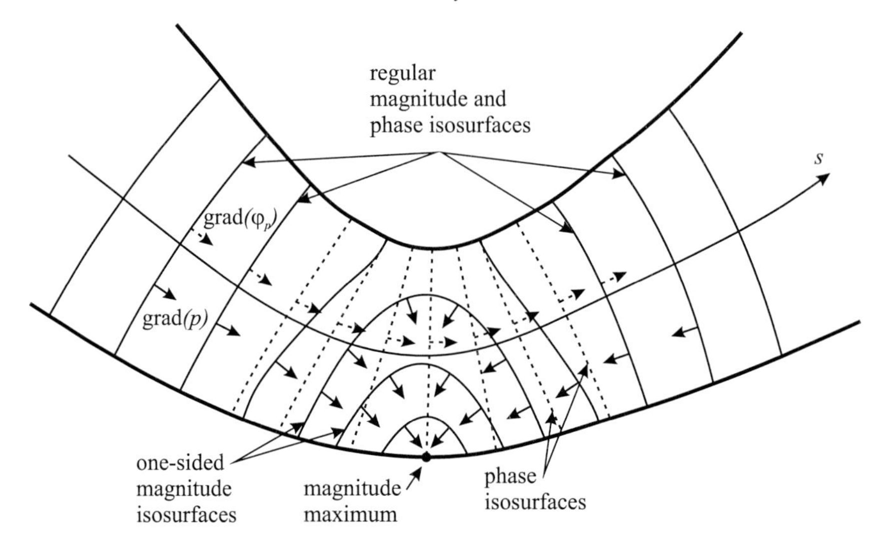

Fig. 27: Schematic representation of the isosurfaces of pressure magnitude (solid lines) and phase (broken lines) and the corresponding gradients in the vicinity of a pressure maximum.

2.4.6 The fundamental sound field of the ear canal

The part of the ear canal sound field that is independent of the external sound source is denoted "core region". Here, the majority of isosurfaces can be characterized as regular. Such surfaces can be used to construct a middle axis for unidimensional models according to the methods listed in section 2.2. In Fig. 3, this process was already illustrated. However, the specification of a middle axis is not strictly feasible, whenever one-sided isosurfaces are present in the canal sound field. In consequence, it can be expected that unidimensional model calculations are subject to errors, in particular, when model ports are specified in ear canal sections with prevalent occurrence of one-sided isosurfaces. However, it can be shown that the transfer characteristics of the core region depend only marginally on the existence of one-sided isosurfaces. In the core region, a given power is transmitted with minimal energy. Due to the slenderness of the canal, forced higher order modes containing additional field energy will convert to fundamental mode forms rapidly after the higher order mode boundary condition is disabled. This feature seems to follow a general principle, as many physical systems enter the state of minimal energy if possible. The condition of minimal energy density is maintained for fundamental modes as well. Thus, it is reasonable to generalize the idea of a "fundamental sound field" in curved acoustical ducts, which is specified in the following.

The ducts considered in the following are supposed to have almost rigid walls. They should have prolate form, so that a wave propagation direction along an approximate middle axis can be specified. The lateral dimensions have to be small compared to the considered wavelengths. Similar to the unidimensional approaches, it is required that the ducts yield exclusively continuous variations of the cross-sectional area and smooth curvature. Further, the requirements of the core region must be met: near-field effects (sources, fluid-structure interactions, scattering or reflections) do not influence the internal sound field.

Stinson, 1985a, describes the acoustical energy flow in ducts. Here, the time averaged rate of acoustical energy flow at a point $\mathbf{r}$ in a duct with plane wave propagation was given as

$$P = \frac{1}{2} A \cdot \mathrm{Re}\left[\underline{p}(\mathbf{r}) \underline{v}(\mathbf{r})^* \right] \qquad (2.4.3)$$

The cross-sectional area of the duct is denoted by A, the complex sound pressure by $\underline{p}(\mathbf{r})$ and the conjugate complex unidimensional velocity by $\underline{v}(\mathbf{r})^*$. Eq. (2.4.3) represents the active power P in the duct. To generalize this equation to surfaces of arbitrary shape (in particular one-sided isosurfaces), we start with a formulation of the energy density W' in a sound field:

$$W'(\mathbf{r}) = \frac{p_{RMS}^2(\mathbf{r})}{\rho c^2} + \rho v_{RMS}^2(\mathbf{r}) \qquad (2.4.4)$$

The orthogonal spatial components of the velocity vector may have arbitrary phase relation (as already discussed in subsection 2.4.5), thus a formulation of the energy density using a magnitude expression exclusively is not sufficient. In Eq. (2.4.4), root-mean-square (RMS) values are used instead of complex amplitudes. According to this term, the local energy becomes minimal for minimal RMS values of sound pressure and velocity.

The complex acoustical power S that is transmitted through a cross-sectional area $\mathbf{A}$ of the duct is calculated by the following integral:

$$\underline{S}(\omega)=\frac{1}{2}\int_A \underline{p}(\mathbf{r})\underline{\mathbf{v}}^*(\mathbf{r})\,d\mathbf{A}=-\int_A \frac{|\underline{p}(\mathbf{r})|^2\cdot\operatorname{grad}\{\varphi_p(\mathbf{r})\}}{2\omega\rho}\,d\mathbf{A}-\mathrm{j}\int_A \frac{|\underline{p}(\mathbf{r})|\cdot\operatorname{grad}\{|\underline{p}(\mathbf{r})|\}}{2\omega\rho}\,d\mathbf{A} \quad (2.4.5)$$

It is assumed that energy is conserved in the duct. Hence, the active power which is identical to the real part of the complex power given by Eq. (2.4.5) is constant in adjacent cross-sections of the duct. The same holds for regular isosurfaces. For plane waves, the real part is identical to Eq. (2.4.3). The power in the duct evaluated over a regular isosurface is represented by Eq. (2.4.5). As the integrals imply a scalar product of the pressure gradients and the area normal, the field components that are normal to the integral surface contribute to the total power exclusively. The tangential components, however, increase the RMS velocity as calculated by Eq. (2.4.4). In conclusion, velocity vectors have to be normal to the pressure magnitude isosurfaces in minimum energy sound fields. This condition is generally met for fundamental modes and regular isosurfaces in the core region (see subsection 2.4.5).

The active power that is transmitted through one-sided isosurfaces has to be zero, because lossless wave propagation is assumed. With $|\underline{p}(\mathbf{r})| = const.$ and Eq. (2.4.5) it follows that

$$\int_A \operatorname{grad}\{\varphi_p(\mathbf{r})\}\,d\mathbf{A}=0 \quad (2.4.6)$$

This equation is fulfilled, if the phase gradient is constant, which is equivalent to uniformly spaced phase isosurfaces. At the exact location of the extremal value that is enclosed by one-sided isosurfaces, the magnitude gradient is zero. Hence, basically the phase gradient determines both orientation and magnitude of the velocity vector. The field in the model duct described in subsection 2.4.5 shows that all phase isosurfaces are regular. Reconsider the time domain velocity term in Eq. (2.4.2). The magnitude gradient which has another orientation than the phase gradient determines the velocity direction only for a short time instant, when the cosine term vanishes. This is fully confirmed by the velocity vector arrows in Fig. 26.

2.4.7 Transfer functions in the ear canal

In subsection 2.4.6 it was shown that the fundamental sound field in the core region follows the principle of minimum field energy even near one-sided isosurfaces. This is also an attribute of fundamental mode propagation. For the desired eardrum pressure estimation method, it is required that transformations over regions containing one-sided isosurfaces yield only minimal error. This issue was analyzed by comparison of sound fields in ducts with exclusively planar wave propagation and in curved cylindrical tubes with the same middle axis length. FE models of semicircular toroidal tubes of 4, 6, 8 and 12 mm diameter and a curvature radius of 15 mm referred to the middle axis of the duct were constructed.

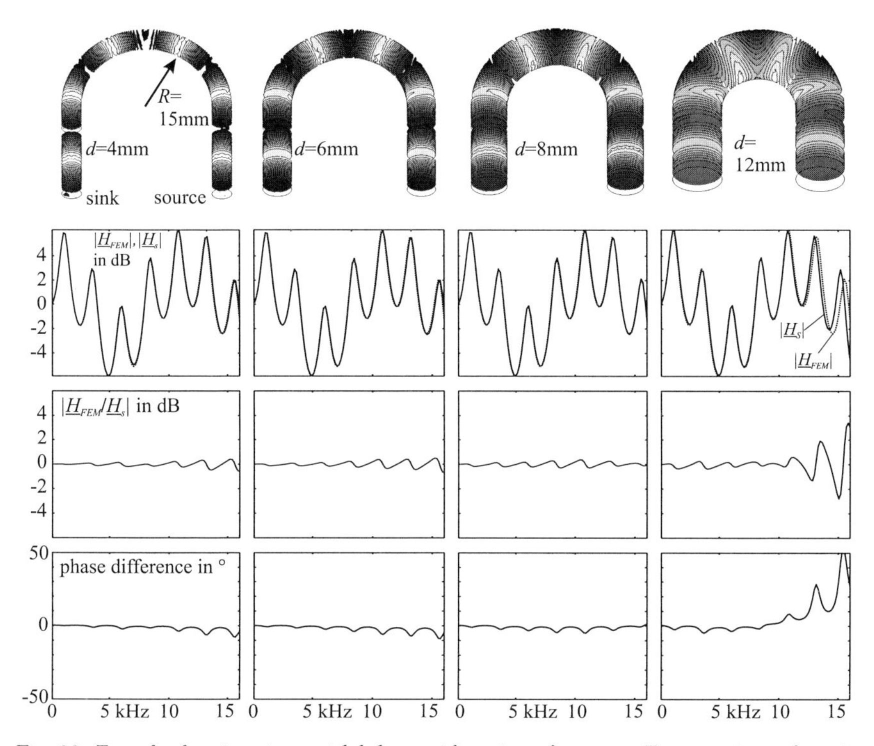

Fig. 28: Transfer functions in toroidal ducts with various diameters. Top row: isosurfaces in the ducts at 12 kHz. Second row: comparison with magnitude transfer functions of straight ducts, third row: difference of the transfer functions, fourth row: phase difference between $\underline{H}_{FEM}$ *and* $\underline{H}_S$

At both ends of the ducts, additional straight tubes were attached, in which the sound fields at the tube ends could continuously merge with the toroid field. The adaptors were chosen long enough to ensure almost perfect plane waves at the connecting surface. The tubes were excited by a volume velocity source at one end. The other end was terminated with an impedance of $Z=2 \cdot Z_{tw}=2 \cdot \rho c/A$. In this configuration, both standing and propagating waves arise in the duct. As expected, one-sided isosurfaces are generated along the toroidal bending (see Fig. 28). The pressure values $\underline{p}_{\text{in}}$ generated near the source (5 mm from the entrance) and $\underline{p}_{\text{out}}$ near the acoustical load (5 mm from the outlet) are taken from the numerical solution to compute the transfer function $\underline{p}_{out}/\underline{p}_{in}$ .The transfer functions are compared to the transfer function of a homogeneous straight tube of the same length.

The deviations are surprisingly small although a number of one-sided isosurfaces appears in the curve. At high frequencies the sound field structure approaches that of the first higher order mode of the toroid. The strongest influence can be observed in the duct with the largest diameter at high frequencies (Fig. 28, right column). In the other cases, the deviations of the transfer functions are not greater than 1 dB in magnitude and 12° in phase angle. Thus, in respect to sound transmission, one-sided isosurfaces are rather marginal disturbances if they are evanescent.

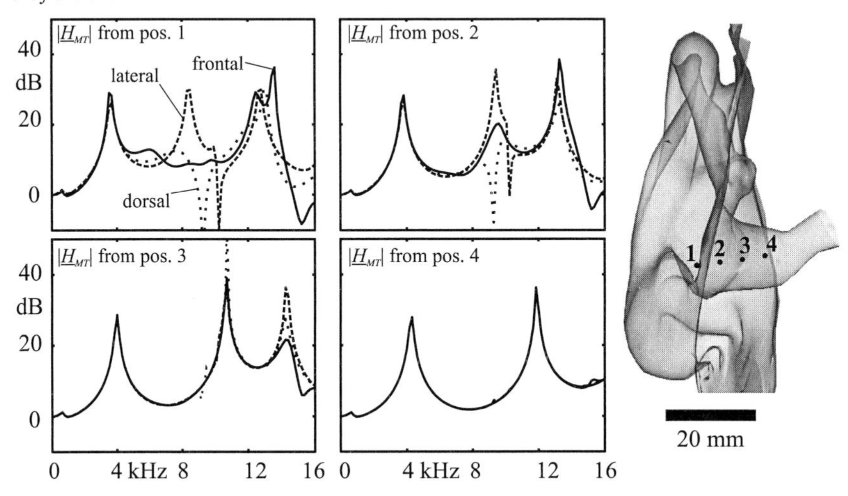

Fig. 29: Point transfer functions from different positions 1-4 near the ear canal entrance to the termination point **T** *at the end of the ear canal. The line styles correspond to the three different kinds of excitation (dorsal, lateral, frontal sound incidence). Right panel: sketch of the pinna with dots indicating positions 1-4.*

Thus, one-sided magnitude isosurfaces that arise in ear canals do not alter transfer functions to the point **T** considerably. However, it is important that the entrance of the ear canal is specified within the fundamental sound field. Otherwise, deviations depending on the external sound field can occur, which can be investigated by examining the effect of the three implemented sources. Pressure transfer functions from four points at the entrance of the canal (refer to the right panel of Fig. 29) to the point **T** have been calculated (left panels of Fig. 29).

The resulting transfer functions for point 1 are distinctly influenced by the source. Differences can be observed already for frequencies above 4 kHz. As expected, the deviations are shifted to higher frequencies, when the measurement position is moved into the canal. At point 2 and point 3, the transfer functions are equivalent up to 8 kHz and 13 kHz, respectively. At these positions, the main errors occur at the transfer function maxima which correspond to pressure minima at the entrance. This indicates the presence of one-sided isosurfaces. The error sensitivity of measurements at locations of pressure minima is further analyzed in the next section. At point 4, the three transfer functions are almost identical. Thus, point 4 can be regarded as "entrance" to the core region of the canal in the considered frequency range.

According to Hammershøi and Møller, 1996, the reference point for HRTF measurements could even be chosen 6 mm outside the ear canal. This clearly disagrees with the presented model calculations. However, the accuracy of the quoted measurements beyond 6 kHz rapidly worsened. Deviations of the order shown in Fig. 29 could hardly be detected.

The experiments with the toroidal duct and the ear canal show that unidimensional transformations of acoustical field variables over irregular regions are hardly altered by one-sided isosurfaces. However, this does not mean that one-sided isosurfaces can be ignored. For instance, in the field inside curved regions, transversal pressure deviations of 10 dB and more can occur, which was not only demonstrated by the described simulation but also reported by other authors (e.g. Stinson and Daigle, 2005). In addition to the simulations, some results of own measurements indicating the influence are documented in subsection 3.2.4.2. The deviation causes essential errors when quantities are related to cross-sectional areas of the ear canal as in the described one-dimensional modeling approaches. In regions with one-sided isosurfaces, the sound field is particularly sensitive to interference with measurement equipment. This issue is discussed in the next section.

2.5 Consequences for measurements at the ear canal entrance

For the determination of $\underline{p}_T$, a calculation of the ear canal transfer function is necessary. By means of one-dimensional approximations, it can be determined from the cross-sectional area function of the canal. As already mentioned in the introduction, inverse methods exist to estimate the cross-sectional area function of straight ducts from their input impedance. The approach by Hudde et al., 1999, was evaluated in artificial inhomogeneous, but straight ear canals. Although it works well in the used canal replicas, it often fails in real ear canals. The simulation results provided in section 2.4 suggest that the errors are generated by the mismatch between the one-dimensional modeling approaches and the three-dimensional sound field. In this section it is examined, how impedance measurements are influenced by spatial sound field distortions at the coupling surface of the measurement device and the ear canal. It can be expected that the entrance of the core region is shifted to more posterior positions in the canal. In this case, the sound field adjacent to the orifice and consequently the measurement results depend on features of the connecting tube.

The acoustical input impedance of the ear canal can be determined in different ways. A very common method is the usage of calibrated sound sources. Norton or Thevenin equivalent sources of the measurement devices that are attached to the ear canal are determined (see subsection 2.5.2). The sources are calibrated using known load impedances. When the source parameters are determined, the impedance at the ear canal entrance can be calculated from a single pressure measurement. Calibrated source measurements are often used in studies of the ear canal (e.g. Møller, 1960; Rabinowitz, 1981; Keefe et al., 1992; Larson et al., 1993; Voss and Allen, 1994; Hudde et al., 1996; Sanborn, 1998). A similar technique is based on the influence of the termination impedance on the pressure transfer functions in ducts ("transfer function method", e.g. Ciric and Hammershøi, 2007. See equation (A.1.18) in Appendix A.1).

Both the calibrated source measurement and the transfer function method require a well-specified coupling of the device outlet and the ear canal. Such a coupling can be realized by individually cast ear molds (e.g. Sanborn, 1998) or foam plugs (e.g. Keefe et al., 1992) which are used to fix insert earphones and microphone probes in the canal. To avoid the area discontinuity between external sound tubes and the ear canal entrance area, the ear canal itself can be used as measurement duct (Hudde, 1983; Voss and Allen, 1994; Farmer-Fedor and Rabbitt, 2002). However, contrary to measurement tubes that can be manufactured to the needs of the particular investigation, the ear canal transmission properties are not known a priori and have to be estimated. In addition, measuring within the narrow ear canal is subject to other types of systematical errors.

When external sound canals are attached to the ear canal entrance, a cross-sectional area discontinuity between the device and the ear canal cannot be avoided. At this interface, higher order modes are excited (Hudde and Letens, 1985; Brass and Locke, 1997; Fletcher et al., 2005; Stinson and Daigle, 2007), but these modes decay at a short distance from the discontinuity. The acoustical waves in the homogeneous connecting tube are planar except very close to the orifice. In contrast, the waves in the ear canal are slightly curved due to the area function and, in particular, due to the curved shape of the ear canal, as was shown in section 2.4.

While impedance minima and maxima of homogeneous tubes have equidistant distribution over frequency, the extrema of ear canal impedances are shifted. This feature is used by the mentioned inverse methods for estimating ear canal area functions. Close approximations of the ear canal transformation are achieved even if only the minimal frequencies are evaluated (Hudde et al., 1999), which can be improved by including the maximal frequencies. In the context of three-dimensional ear canal sound fields, is important to examine the impact of the coupling of the connecting tube and the ear canal for minima and as well for maxima arising at the entrance.

2.5.1 Models of ear canal impedance measurements

The spatial sound field that arises when a measurement device is connected to the ear canal was examined using an FE model basing on the components of the external ear simulations introduced in section 2.3. Ear canal, tympanic membrane and middle ear were separated from the design and connecting tubes were attached. Again, the walls of the connecting tube and of the ear canal are assumed to be rigid and the sound propagation in both ducts is modelled lossless. Although wall vibrations and losses affect the sound propagation in natural ear canals, general characteristics of the spatial field are modelled sufficiently realistic. The geometry of ear canal, tympanic membrane and middle ear is kept constant throughout the calculations. The models only differ with respect to shape and alignment of the connecting tube (Fig. 30). The effect of the area discontinuity is examined by three variations of the tube diameter. Two of the tubes are implemented as circular ducts which have diameters of 2 mm ("thin") and 4 mm ("medium"). The third model provides a flush connection between the connecting tube and the ear canal. For that purpose, the entrance area of the ear canal model is extruded along its normal. By this means a maximally smooth transition is constructed. The extruded tube is not very realistic, but is regarded as "ideal" device because field disturbances at the interface are reduced in comparison to the tubes with area discontinuities.

The cross-section of the extruded canal corresponds to a circular tube diameter of approximately 9.3 mm. The position of the coupling area in reference to the ear canal is the same in each of the three cases. At first, all three connecting tubes are aligned perpendicular to the cross-sectional area at the ear canal entrance. The ear canal entrance area was selected considering the isosurface plots shown in section 2.4. It is located near the entrance of the core region arising in the free ear canal. Hence, the orifice of the connecting tube is approximately parallel to the local isosurfaces. In practice, however, another systematical error arises in addition. It is not easy to place the orifice according to the a priori unknown isosurfaces of the subject's ear canal. Often the connecting tube cannot be aligned correctly with the orientation of the ear canal because parts of the pinna (in particular the *tragus*) obstruct the direct access. Then the tube axis differs from the normal vector of the ear canal input area at the location of the tube orifice. Hence, the alignment of the connecting tube is not at all unique.

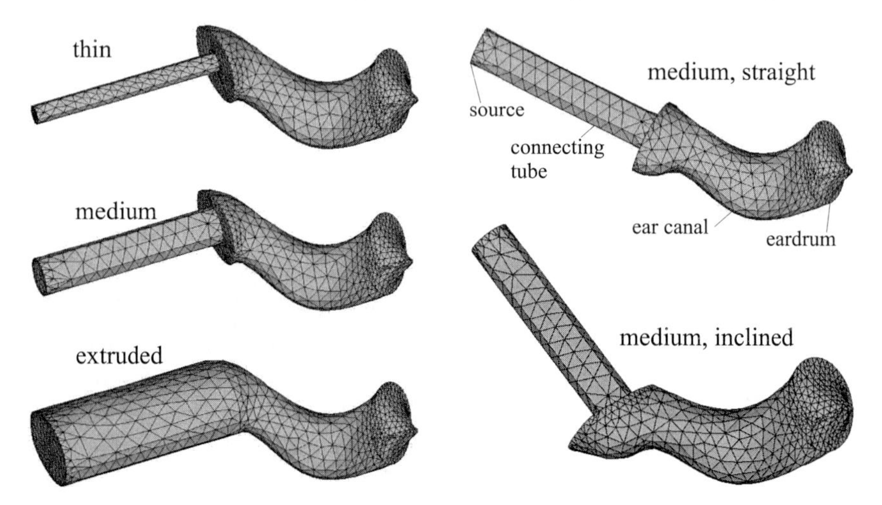

Fig. 30: Investigated impedance measurement models. The connecting tubes have different diameters or different adjustments with respect to the ear canal. The middle ear model is not shown in the plots.

To examine the influence of its orientation, two models with inclined connecting tubes were generated, however exclusively for the medium diameter tube because it is the most practicable. The tube was rotated about the center of its orifice area, which keeps the resulting ear canal length constant. As the "original" medium tube as given in Fig. 30 (left column) exhibits a significant inclination with respect to the axis of the foremost part of the ear canal, one of the two variations is constructed aiming at a straight continuation ("medium, straight") of this axis. The "medium, inclined" case is another variation which is rotated by 30 degrees referred to the original position.

In the FE models, the mean nodal distance of the mesh is generally adjusted to 1.5 mm (approximately 15 elements per wavelength at 16 kHz). Near the coupling tube orifice, the spatial variation of the sound fields at the interface is much stronger than at some distance, particularly in the case of large area discontinuities. Here the nodal distance needs to be increased. By increasing the node density until no significant differences in the calculated sound fields were observed, the necessary nodal distance was determined to be 0.7 mm.

The sound field inside the models is excited by sources at the free end of the measurment ducts. If the sources were rigid, strong transmission line resonances would be obtained because the tympanic membrane is highly reflecting as well. Such resonances can degrade the numerical accuracy of the computations. Therefore a fully absorbing sound source at a sufficient distance from the orifice of the tube is simulated using a vibrating cross-sectional surface ("piston") of impedance equivalent to the wave impedance $Z_w=\rho c$. In real measurements damping material, placed near the driving transducer, is necessary as well to avoid increasing errors near resonances. Frequencies are chosen in equidistant steps of 160 Hz up to 16 kHz

2.5.2 Simulation results

The impedance at the orifice of the connecting tube in the FE model is "measured" using the calibrated source method. The pressure excited by the implemented source allows the load impedance to be calculated. Fig. 31 shows the corresponding Norton equivalent circuit.

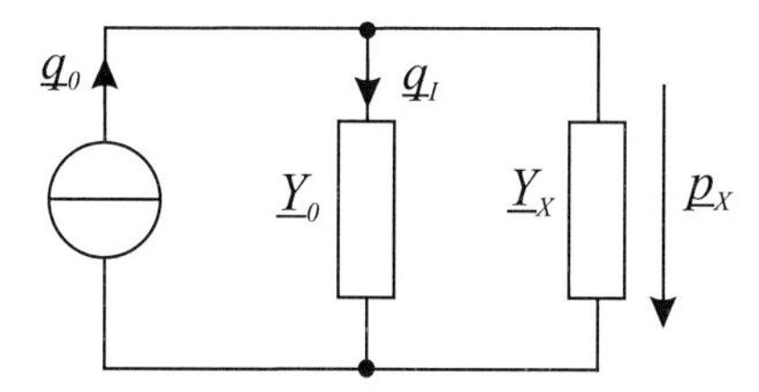

Fig. 31: Norton equivalent circuit representing a volume velocity source $\underline{q}_0$ having an internal admittance $\underline{Y}_0$. When the source parameters are known, the load admittance $\underline{Y}_x$ can be calculated from the pressure $\underline{p}_x$.

For a volume velocity source $\underline{q}_0$ having an internal admittance $\underline{Y}_0$ the load admittance $\underline{Y}_x$ is calculated from the "measured" pressure $\underline{p}_x$ as

$$\underline{Y}_x = \frac{\underline{q}_0}{\underline{p}_x} - \underline{Y}_0 \qquad (2.4.7)$$

In the modeled cases, the internal admittance is represented by the tube wave admittance $Y_0 = Y_{tw} = A / Z_w = A/(\rho c)$ where A denotes the cross section of tube and piston. The volume velocity of the source is derived from the applied piston displacement $\underline{\xi}$ according to $\underline{q}_0=\mathrm{j}\omega\underline{\xi}A$. The pressure $\underline{p}_x$ is determined at the central node of the piston which can be considered as virtual microphone position. To calculate the impedance $\underline{Z}_{out}$ at the orifice of the connecting tube, the admittance $\underline{Y}_x$ has to be transformed according to

$$\underline{Z}_{out} = Z_{tw} \frac{\cos(\beta L_t) - \mathrm{j}\underline{Y}_x Z_{tw} \sin(\beta L_t)}{\underline{Y}_x Z_{tw} \cos(\beta L_t) - \mathrm{j}\sin(\beta L_t)} \quad (2.4.8)$$

The product βL_t of wave number and tube length is perfectly known because the speed of sound and the dimension of the tube are specified as model parameters. The lossless transformation underlying Eq. (2.4.8) accurately matches the conditions simulated. Consequently, it is expected that the impedance at the orifice of the measurement tube equals the input impedance of the ear canal, which is the same for each of the simulated cases. Impedances "measured" under this assumption are represented in Fig. 2. The impedances are normalized to the tube wave impedance of a cylindrical duct of 8 mm diameter. Obviously the utilized connecting tubes have considerable impact on the "measured" impedance. The most striking effect of the area discontinuity can be observed in impedance minima. There is a systematical shift of measured minimum frequencies to lower values if the cross section of the connecting tube is reduced. In contrast, the impedance maxima are essentially less affected.

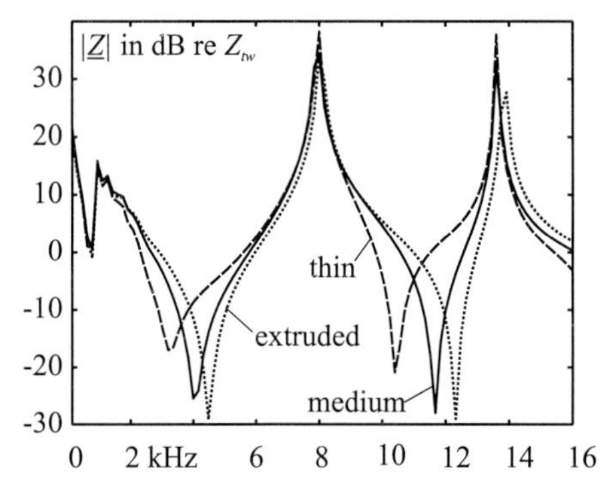

Fig. 32: Ear canal impedances "measured" by simulation of three different connecting tubes (cases "thin", "medium", "extruded" according to Fig. 30). The dip at low frequencies is caused by resonances of the tympanic membrane and the middle ear (compare subsection 2.4.2).

At the orifice, spatial sound field structures arise which make the ear canal appear a little longer than it is. The decrease of characteristic frequencies of tubes coupled to a general sound field is well known. It can be basically interpreted as an effect of the additional mass directly behind the tube orifice that vibrates with the sound field in the tube. In some cases, mouth corrections are possible which approximate the three-dimensional effect in terms of the one-dimensional model, of course, only for sufficiently low frequencies. For the special case of an area discontinuity occurring in a circular cylindrical duct an approximate formula for the acoustical mass can be used (Karal, 1953).

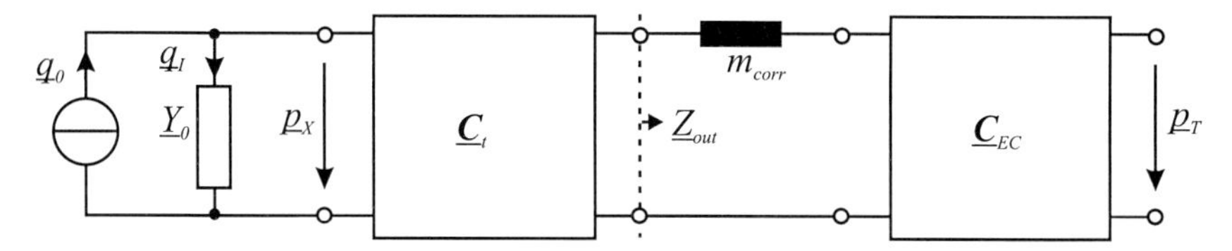

Fig. 33: Schematic with added equivalent mass m_{corr} representing spatial sound field effects at the area discontinuity between the connecting tube (represented by the two-port $\underline{C}_t$) and the ear canal ($\underline{C}_{EC}$).

Karal's approximation provides a good estimate of the effective mass except for small area ratios at the discontinuity (Hudde and Letens, 1985). Thus, Karal's correction can be expected to improve the measured impedances.

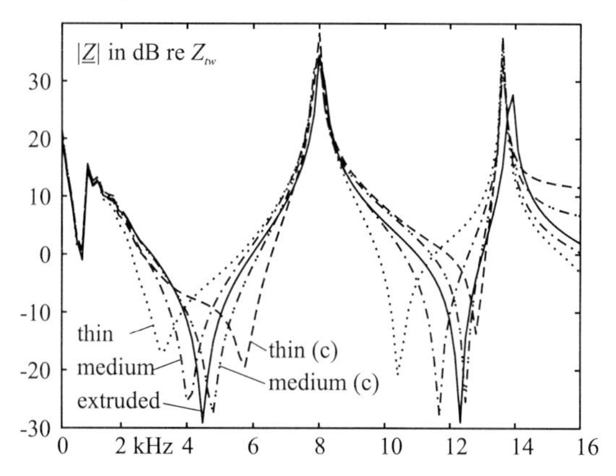

Fig. 34: Ear canal impedances "measured" by simulation of three different connecting tubes (cases "thin", "medium", "extruded" according to Fig. 30) and corrected using Karal's mass impedance at the discontinuity. Corrected impedances are indicated by (c). For the extruded tube no correction is applied because no discontinuity exists.

According to the corresponding equivalent circuit (Fig. 33), Karal's acoustical mass impedance simply adds to the actual input impedance of the ear canal. Therefore the mass impedance is subtracted to obtain the corrected versions of the impedances that were given in Fig. 32. It can be expected that the extruded case will give the best approximation of the correct impedance because an area discontinuity is completely avoided (consequently, no correction mass is applied).

The corrected results are disappointing: the minima are shifted in excess of the frequency calculated for the extruded tube, which is regarded as correct frequency here (Fig. 34). For the depicted cases, half of Karal's mass would yield better results. The results show that the correct mass is critically depending on the shape of the ear canal input area and on the area function and curvature of the section just behind the interface. This was examined using an FE model of the experimental duct used for measurements in Hudde and Letens, 1985. In the simulation, a system of two circular cylindrical ducts (length 50 and 110 mm and diameter 15 and 50 mm, respectively) is excited at the free end of the smaller tube. The impedance at the area discontinuity is determined according to the method described above. To analyze the influence of the termination, two cases with varying boundary impedance at the far end of the larger tube are implemented (rigid termination and termination with the tube wave impedance to achieve anechoic conditions). Karal's correction is restricted to circular ducts and to wavelengths that are long in comparison to the tube radii. The characteristics of ducts that are not axially symmetric were examined using a further model. Here, the large tube has square cross section; the surface area is equal referred to the first case (Fig. 35, left column).

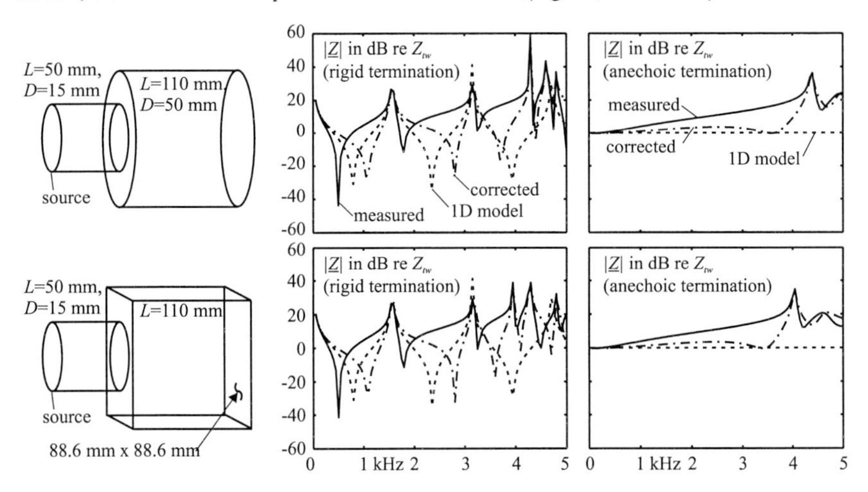

Fig. 35: Impedances "measured" in circular and rectangular ducts for rigid and matched termination (FE simulation).

Similar to the experiment documented by Hudde and Letens, the frequencies of the simulation range from 0 to 5 kHz in 100 equidistant steps. A reference impedance for comparison can be determined analytically using a one-dimensional model. The results of the impedance calculations are depicted in Fig. 35 as well. Similar to the ear canal impedances in Fig. 34, the impedance minima of the rigidly terminated ducts are not compensated correctly to match the analytic solution (center column). For anechoic termination, however, Karal's correction significantly improves the impedance results (right column), at least up to the first characteristic frequency of the simulated duct at 3-4 kHz. The results of the circular and rectangular duct (top and bottom row, respectively) are very similar up to the mentioned frequency. Obviously, in both ducts a fundamental mode field is present. For higher frequencies, the sound field shape in the larger ducts is significantly different. Apparently, Karal's approximation is not useful, when the exact frequencies of impedance extrema are to be determined.

The calculations can also be used "inversely" to determine the difference between the simulated and expected impedance. By this means, the mass that becomes acoustically effective can be compared using the prediction of Karal's equation. For that purpose, the analytically calculated impedance is subtracted from the measured values. In Fig. 36, the imaginary parts of the results are shown (in the fundamental mode range below 4 kHz, the impedance is purely imaginary).

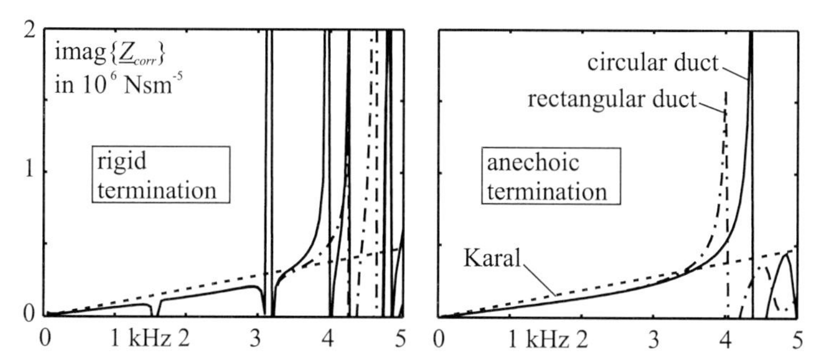

Fig. 36: Imaginary part of the additional impedance represented by the corrrection mass.

In the case of anechoic termination (right panel), Karal's solution approximates the mass effect fairly well below 3 kHz. Here, the measurements of Hudde and Letens are reproduced in good approximation. However, in the impedance extrema that occur in the rigidly terminated duct (left panel), the mass model is not compatible. This suggests that no impedance extrema occurred in the frequency range of the measurements. Unfortunately, the correction has to be most accurate particularly at the extrema.

In conclusion, no simple solution can be found to estimate the mass correction more accurately. Regarding the equivalent circuit in Fig. 33, the small additional mass essentially alters the magnitude of impedance minima, whereas the impedance magnitude in maxima remains nearly unchanged, hence the mass correction has only small impact on maxima. This can be found in the impedance plots of Fig. 34 as well: the maxima are constant, although the mass correction changes the results significantly in a relatively broad frequency range around the minima. This concerns even impedance magnitudes higher than the tube wave impedance, which is represented as 0 dB in Fig. 34.

In addition to the extension effect that alters the position of the minima, impedances are affected by the misalignment of the tube. The corresponding impedance curves are depicted in Fig. 37. The results were calculated using the medium, inclined and straight connection tubes (Fig. 30). The figure suggests that the tube orientation has a generally smaller effect than the area discontinuity. In contrast to the first part of the investigation, the minima remain practically unchanged; the major difference arises at the maximal frequencies.

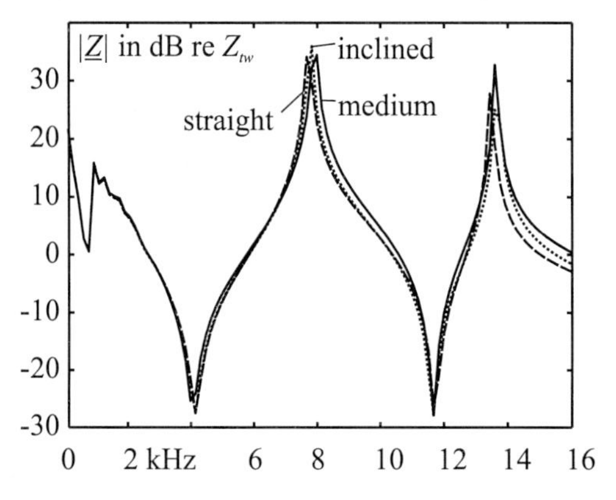

Fig. 37: Ear canal impedances "measured" by simulation of three different coupling angles using the medium tube (cases "medium", "straight", "inclined" according to Fig. 30).

The FE models allow for an investigation of the isosurface shape near the orifice of the connecting tube. In the following the impact of the tube shape and orientation is investigated by considering the sound field structure.

2.5.3 Sound field structure at the interface

The connecting tube and the ear canal represent an acoustical duct system. The internal sound field is a superposition of standing and propagating waves. Hence, minima and maxima of the sound pressures and velocities alternate (At the position of pressure maxima, volume velocity minima arise, and vice versa). The position of the extrema depends on frequency. It is reasonable to examine the sound field at the minimal and maximal frequencies of the impedance magnitude, because the impedances show major deviations particularly at those frequencies. As in previous sections, the field structure at the orifice is visualized by magnitude isosurfaces and velocity vector arrows. Again, the pressure and velocity amplitudes are separately normalized for each case and the dynamic range $D=20\cdot\log_{10}|p_{max}/p_{min}|$ of the pressure field is given in the figures. The pressure range is divided into 20 equidistant steps.

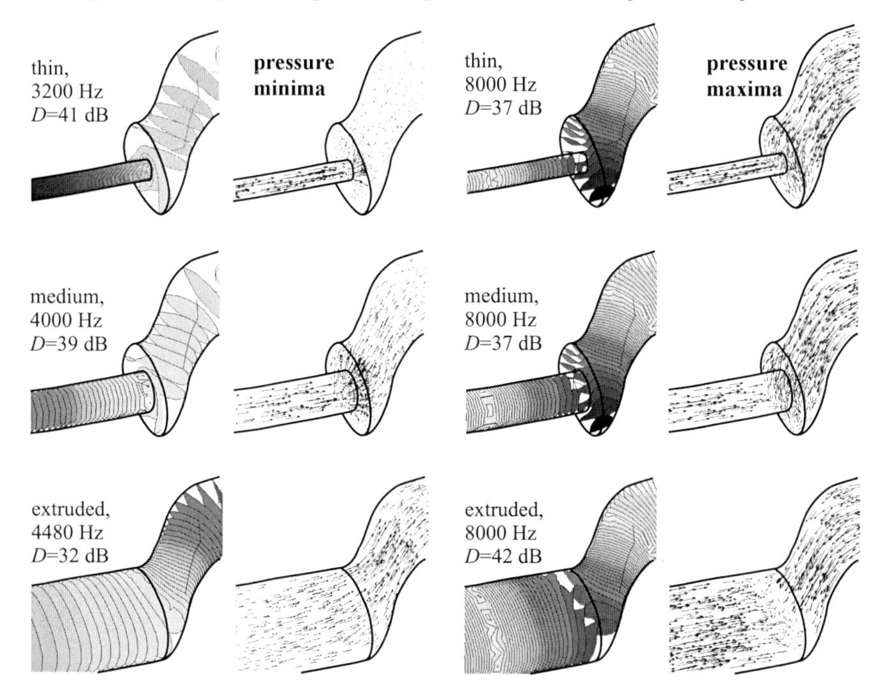

Fig. 38: Isosurfaces of pressure magnitude and instantaneous velocity vectors calculated for different connecting tubes in the case of pressure minimum (left) or maximum (right) at the interface between the tube and the ear canal. The frequency values denote the extremal frequencies within the resolution applied in the finite element calculations (multiples of 160 Hz).

Fig. 38 and Fig. 39 depict the sound fields that arise for different area discontinuities and tube adjustments, when the first pressure minimum (left column) and the first maximum (right column) appear at the interface area. For the plots, the frequency closest to the impedance extrema as determined from Fig. 32 and Fig. 37 was selected from the set of frequencies applied in the FE simulation. Obviously, distinctly three-dimensional isosurfaces can be found.

The frequency of the first pressure minimum ranges from 3200 Hz for the thin connecting tube up to 4480 Hz for the extruded tube (Fig. 38, left; compare Fig. 32). As expected the isosurfaces close to area discontinuities strongly deviate from regular, almost planar shapes that arise in the posterior part of the ear canal and especially in the connecting tubes. The isosurfaces obtained for the extruded tube also exhibit some change during transition to the ear canal, but appear widely regular.

The pressure minimum is equivalent to a local velocity maximum. The ear canal has an acoustical length of approximately a quarter wavelength; hence its high termination impedance is transformed into low input impedance. Both pressure isosurfaces and velocity vectors show that the sound transmission from the connection tube has the character of radiation into the ear canal. The velocity vectors at the orifice spread radially into the space behind the discontinuity. After a small distance, the vectors rearrange and form the regular ear canal field, where the vectors are widely directed parallel to the walls again. The normal component of the particle velocity disappears on the rigid device boundary area surrounding the port. The velocity flow lines which run close to the ear canal walls are considerably curved just behind the interface, therefore the traveling distance of sound waves is longer than the middle axis through tube and ear canal. Thus the mean distance between the tube orifice and the termination of the ear canal is extended compared to a tube without discontinuity. The ear canal appears longer than it is. Consequently, the minima are shifted to lower frequencies. The detour becomes longer if the discontinuity increases, i.e., if the tube diameter decreases. If the cross-sectional area of the connecting tube is identical to the entrance area of the ear canal, no extension effect takes place. Therefore the extruded tube produces impedance results which are close to the correct input impedance. However, individually matched extruded connection tubes are hardly feasible in practice.

For each of the three cases, the first pressure maximum at the connecting tube orifice occurs for approximately 8 kHz. The corresponding isosurface and vector arrow plots can be found in Fig. 38, right column. In terms of one-dimensional approaches, the ear canal is half a wavelength long and thus transforms the high termination impedance to the entrance. The spatial sound field structures of maxima and minima are fundamentally different. The sound fields in the connecting tube and in the ear canal are decoupled, i.e. isosurfaces and vector arrows in the two regions are orientated differently. In contrast, when a pressure minimum occurs at the interface, the shape of isosurfaces continuously merges from the tube to the ear canal and the velocity vectors have the same orientation. As the pressure maximum at 8 kHz corresponds to a velocity minimum at the same location, the velocity vectors on both sides of the interface are in opposite phase. As the tympanic membrane reflects incident waves almost totally, a high standing wave ratio arises in the canal. Hence, the magnitude of velocity is almost zero in the minimum. The sound fields in the connecting tube and in the canal are thus decoupled, when a pressure maximum arises. The orientations of velocities in the two regions are then independent of each other. In contrast, a pressure minimum with high velocity magnitudes robustly couples both fields. Although one-sided pressure isosurfaces appear directly at the tube orifice in the extruded case, the measurement error is small due to the weak velocity coupling.

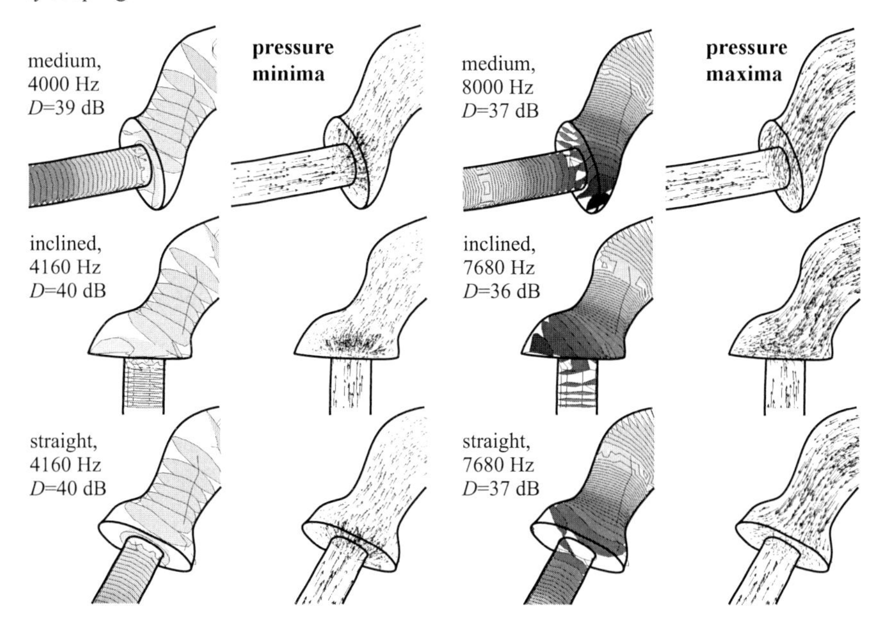

Fig. 39: Pressure isosurfaces and velocity vectors for connecting tube angle variation. Left: pressure minimum at the port area, right: pressure maximum at the port area. The frequency values denote the extremal frequencies within the resolution applied in the finite element calculations (multiples of 160 Hz).

As the isosurfaces are always aligned perpendicular to the rigid walls, the surface surrounding the interface area guides the pressure waves. In the direct vicinity of the interface area, the two fields merge and are thus weakly coupled. Here, a distinct break in the isosurfaces can be found. The canal isosurfaces protrude into the orifice of the connection tube. Hence, wave fronts at the end of the tube are not strictly planar. As the isosurfaces are not spherical like in Fig. 38, left column, the apparent length of the ear canal is not changed. It can easily be seen by the darkest part of the isosurfaces that the exact location of the maximum is identical in all three cases. It is determined by the shape of the ear canal near the interface. As a result, the frequencies of the maxima are the same.

In contrast, the maximum frequencies are affected by the orientation of the connecting tube with respect to the ear canal, as seen in Fig. 37. To achieve an approximately constant effective length of the ear canal, the tube of the "medium" case was rotated around the center of its orifice. Thus, the cases "inclined" and "straight" were obtained. According to the tube inclination, the position of the orifice plane is varied as well. Consequently, the pressure maxima vary their position to some extent. The resulting sound field geometry is visualized in Fig. 39, right column.

The corresponding shift of minimal frequencies is small compared to the effect of the area discontinuity (Fig. 39, left column). As the tubes in the three cases have identical cross section, the sound fields obtain equal apparent ear canal lengths. The radially spreading structure of velocity vector arrows follows the orientation of the interface plane. Additionally, the field structure is less variable due to the strong coupling in velocity maxima.

In conclusion, the frequency of impedance minima is basically influenced by the area step between the measurement device and the ear canal, whereas the maximal frequencies are affected most by the angle between the two ducts. The influences are, however, not strictly separable. Both effects are minimized by the extruded case (see figure 2, bottom) as no discontinuity occurs and a smooth transition between the canals is obtained. Unfortunately, it is very impractical to manufacture extruded connecting tubes for natural ear canal geometries, in particular for large numbers of subjects.

In addition, extruded tubes do not provide perfect coupling as well. An ideal connecting tube should preserve the shape of isosurfaces that would arise in the free ear canal. As a consequence, its orifice must be adapted to the entrance of the fundamental sound field of the ear canal. Here, the field is by definition independent of the external source. Local regular isosurfaces can be interpreted as entrance port of the ear canal and as a reference area for the impedance measurement. The connecting tube, however, must not distort the local fundamental sound field. It has to continue the ear canal outwards without changing the orientation or shape of isosurfaces at the interface. Consequently, the orifice must be adapted to the isosurface at the orifice, i.e. generate wavefronts that perfectly match it. As the form of isosurfaces may significantly depend on frequency, such tubes are hardly feasible. Furthermore, the tube must form a smooth continuation of the ear canal without discontinuities in cross-sectional area function and middle axis.

In conclusion, it is not reasonable to estimate the eardrum pressure using models that are developed from impedance data measured at the ear canal entrance. The exact frequency of impedance extrema is determined essentially by the shape of the ear canal. Hence, it is necessary to determine the minima and maxima very accurately to derive the desired one-dimensional model from the geometry. The examples provided in this section show that it is not possible to detect the particular frequencies with high precision. In general, significant measurements of the ear canal entrance impedance are questionable for frequencies beyond 3-4 kHz. Below this frequency, connecting tube and ear canal can be regarded as lumped element system. No wave propagation effects and consequently no minima arise at the interface. This allows for diagnostic measurements up to 3 kHz (e.g. tympanometry). For higher frequencies, the tympanic membrane can be assumed to be almost rigid. Thus, the ear canal impedance in the frequency range above 3-4 kHz is essentially influenced by the shape of the ear canal, while the eardrum influence can be neglected. For higher frequencies, the determination of eardrum impedances by transformation from the canal entrance is generally very inaccurate (Hudde and Engel, 1998a).

3 Measurements with relation to the eardrum sound pressure

The findings documented in section 2.5 suggest that it is not reasonable to develop an estimation approach for the eardrum pressure signal $\underline{p}_T$ basing on impedance measurements at the ear canal entrance. On the one hand, frequencies of impedance extrema strongly depend on the coupling of measurement devices that are attached directly to the ear canal. On the other hand, the sound field in the ear canal generally shows spatial structures which are incompatible with the underlying one-dimensional modeling approaches. However, it was shown in section 2.4 that pressure transfer functions between points in the fundamental sound field (and in particular from the ear canal entrance to the point **T**) remain meaningful. This notion suggests a new eardrum signal estimation approach which is introduced in the following sections (Fig. 40).

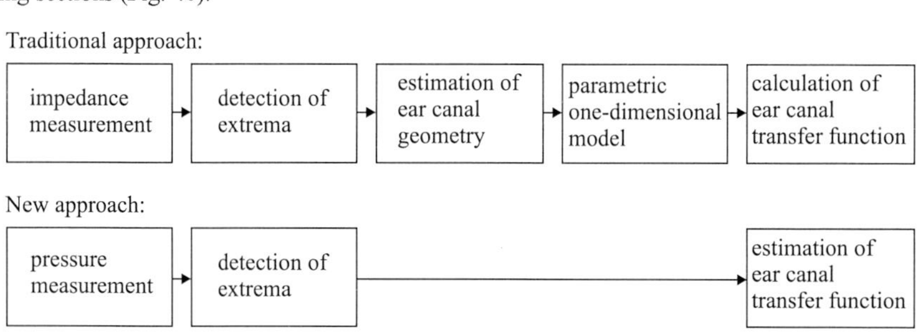

Fig. 40: Schematic of the eardrum pressure estimation approach introduced in the following sections in comparison to state-of-the art approaches.

Instead of developing a one-dimensional model from the estimated ear canal geometry prior to the calculation of the transfer function, the pressure transfer function between the pressure measurement position and the tympanic membrane is estimated directly. The estimation is based on the pressure signal occurring at an arbitrary point in the fundamental sound field. This approach yields two essential advantages:

- The use of an auxiliary one-dimensional model which is per definition not strictly compatible with the sound field in the ear canal is avoided.
- Instead of measurements with connection tubes attached to the ear canal, point measurements with miniature microphones are carried out.

As already mentioned in the introduction (section 1.2), similar methods are described in literature. In some references, the residual length of the canal is estimated by the equivalent acoustical compliance that results when the air volume enclosed in the canal is considered and the canal diameter (e.g. Larson et al., 1993). Although these methods avoid impedance measurements, they are restricted to low frequencies that allow a lumped element approximation of the ear canal. Estimations are often based on the first minimum of pressure or impedance (Chan and Geisler, 1990; Siegel, 1994; Huang et al., 2000; Storey and Dillon, 2001). Other authors compensate several minima to increase the accuracy (Stevens et al., 1987). However, a relatively long measurement tube which is attached to the canal has to be used for this technique. It underlies the disadvantages that occur, whenever an external measurement tube is attached to the ear canal (section 2.5).

The proposed method implements broadband pressure measurements using probe microphones in the canal and works on the first two pressure minima. Thus, the estimation accuracy for higher frequencies is greatly enhanced compared to methods that compensate only the first minimum, as well as coupling effects between the canal and an attached connection tube of a measurement device are avoided, because probe microphone measurements in the free ear canal are applied. It turned out that it is sufficient to determine signals using only one probe in the fundamental field to achieve accurate results.

In the following, the estimation method for the transfer function between the measurement position and the tympanic membrane is described in detail (section 3.1). The method was verified using FE simulations and measurements in an artificial ear (section 3.2). As first application results, measurements of equal-loudness level contours with reference to the eardrum pressure are presented (section 3.3).

3.1 A method for the estimation of the eardrum signal

Basically, the following steps are necessary to determine the eardrum pressure $\underline{p}_T$ from signals at the ear canal entrance:

1. Measurement of a pressure signal at a point in the fundamental sound field of the ear canal.
2. Identification of the first two pressure minima that originate from duct effects. The corresponding frequencies are determined with high precision.
3. A pressure transfer function that compensates the detected minima has to be determined.
4. The initially measured pressure signal is transformed using the resulting transfer function.

The next section focuses on the necessary pressure transfer function. It is derived from a duct model which is only used as adaptive equivalent system and must not be interpreted as approximation for the residual ear canal geometry. Practical approaches for identifying the required minima in the ear canal pressure and for adjusting the necessary model damping are discussed. The measurement system that was developed as practical implementation of the method is documented afterwards.

3.1.1 Transfer functions for minimum compensation

The equivalent duct is modelled using a one-dimensional approach. Sound pressure p and volume velocity q can be transformed between two positions denoted 1 and 2 by a set of transmission equations, which are formulated as chain matrix C:

$$\begin{pmatrix} \underline{p}_1 \\ \underline{q}_1 \end{pmatrix} = \underline{\mathbf{C}} \begin{pmatrix} \underline{p}_2 \\ \underline{q}_2 \end{pmatrix} = \begin{pmatrix} \underline{C}_{11} & \underline{C}_{12} \\ \underline{C}_{21} & \underline{C}_{22} \end{pmatrix} \begin{pmatrix} \underline{p}_2 \\ \underline{q}_2 \end{pmatrix} \qquad (3.1.1)$$

The corresponding pressure or volume velocity transfer functions depend on the chain parameters $\underline{C}_{mn}$ and on the acoustical impedance $\underline{p}_2/\underline{q}_2$ seen at the transformation target (a basic deduction of the equation system and the calculation of duct transfer functions is given in Appendix A.1). When the termination of the duct is modeled to be nearly rigid (which is given ear canals, see sections 2.4.2 and 3.1.4), the point transfer function $\underline{H}_{MT}=\underline{p}_T/\underline{p}_M$ between the measurement position **M** in the sound field of the duct and its termination point **T** is essentially determined by the chain parameter $\underline{C}_{11}$:

$$\underline{p}_M = \underline{C}_{11}\underline{p}_T + \underline{C}_{12}\underline{q}_T \cong \underline{C}_{11}\underline{p}_T \rightarrow \underline{H}_{MT} = (\underline{C}_{11})^{-1} \quad (3.1.2)$$

The spectrum of the entrance pressure $\underline{p}_M$ shows distinct minima that depend on the distance to the point **T** and on the duct shape. To match the transfer characteristics of the equivalent duct and the given residual ear canal, the chain parameter $\underline{C}_{11}$ of the model duct is adapted by fitting its two least minimal frequencies to those occurring in the microphone pressure. With the resulting chain matrix, an extremely accurate transfer function estimate in the considered frequency range is obtained. The adjustable shape parameters of the equivalent duct model are its length and the ratio of its termination and entrance radii. Fig. 41 shows the definition of the parameters (duct length L, entrance radius r_1, and termination radius r_2) and the resulting two-port equivalent.

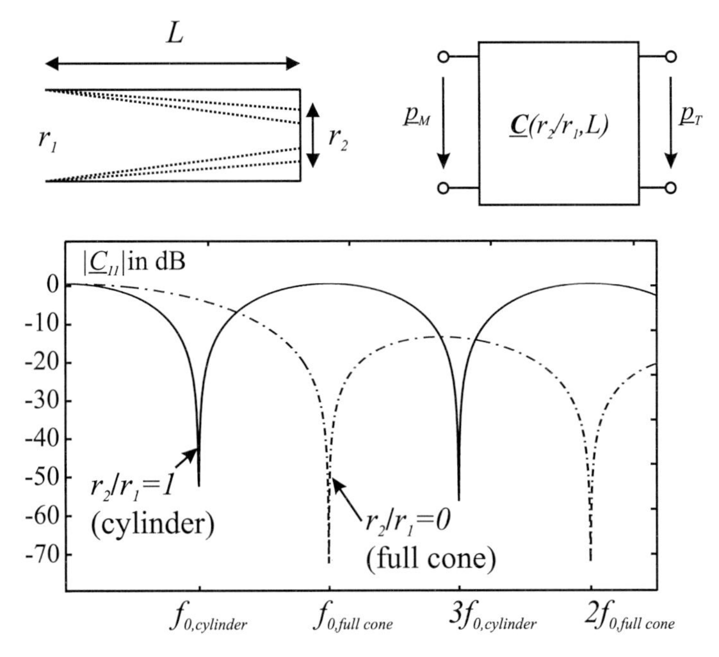

Fig. 41: Top row, left: the shape parameters r_2/r_1 and L of the duct model are adapted to fit the minima of the measured sound pressure. Right: two-port equivalent with specified input and output pressure and rigid termination (open port). Bottom panel: chain parameter $\underline{C}_{11}$ for maximal variation of r_2/r_1 with constant L.

Using the two parameters, the entrance impedance of the duct can be varied between the case of a cylindrical tube yielding minima at odd multiples of the first minimal frequency and the case of a full conical duct with minima at all integer multiples. The duct chain parameter $\underline{C}_{11}(r_2/r_1,L)$ results to

$$\underline{C}_{11} = \frac{r_2}{r_1}\cos(\beta L) - \left(\frac{r_2}{r_1} - 1\right)\frac{\sin(\beta L)}{\beta L} \tag{3.1.3}$$

with $\beta = \omega / c$ denoting the wave number. Obviously, $\underline{C}_{11}$ does not depend on absolute values of r_1 and r_2, but exclusively on their ratio. For two given minimal frequencies with corresponding wave numbers β_1 and β_2, equation (3.1.3) can be separated in the following way:

$$\frac{r_2 / r_1}{r_2 / r_1 - 1} = \frac{\tan(\beta_1 L)}{\beta_1 L} = \frac{\tan(\beta_2 L)}{\beta_2 L} \tag{3.1.4}$$

The length parameter has to be calculated by solving the transcendent equation given by the two terms on the right. After that, the ratio of r_2 and r_1 can be obtained by solving the term on the left. One of the measured minimal frequencies can be selected arbitrarily by variation of the microphone position. It is, however, crucial to place the microphone in the core region of the canal, where the sound field structure does not depend on the source. The presence of one-sided isosurfaces is not critical. Estimations carried out with the toroidal model mentioned in subsection 2.4.7 show that accurate results are obtained even if the microphone is located at one-sided isosurfaces in duct bendings. However, transfer functions from two different points with lateral distance to the tympanic membrane may vary significantly, even if the distance between the points is extremely small. If only the first minimum is regarded, a basic transformation is obtained by using $\underline{C}_{11} = \cos(\beta L)$ with $L=c/4f$ calculated from the first minimal frequency. The equivalent cone model then becomes an equivalent cylindrical duct. Similar techniques are described in literature (some references are listed in the introduction of this chapter). Approximations basing on two minima and both geometry parameters r_2/r_1 and L will be denoted *second-order estimation* in the following, whereas those including only the first minimum are referred to as *first-order estimations*. It is important to underline that the resulting model does not represent a geometrical approximation of the residual ear canal. In particular, the length parameter L is not identical with the distance between the microphone and the point **T**, as the canal shape influences the minimal frequencies as well.

3.1.2 Detection of minima and model fitting

To achieve measurement data, a probe microphone is inserted into the ear canal, which is excited by a broadband sound signal (white noise). The estimation method relies on the precise detection of minima that arise due to destructive interferences of the incident and reflected waves in the ear canal (in the following, these are referred to as "standing wave minima").

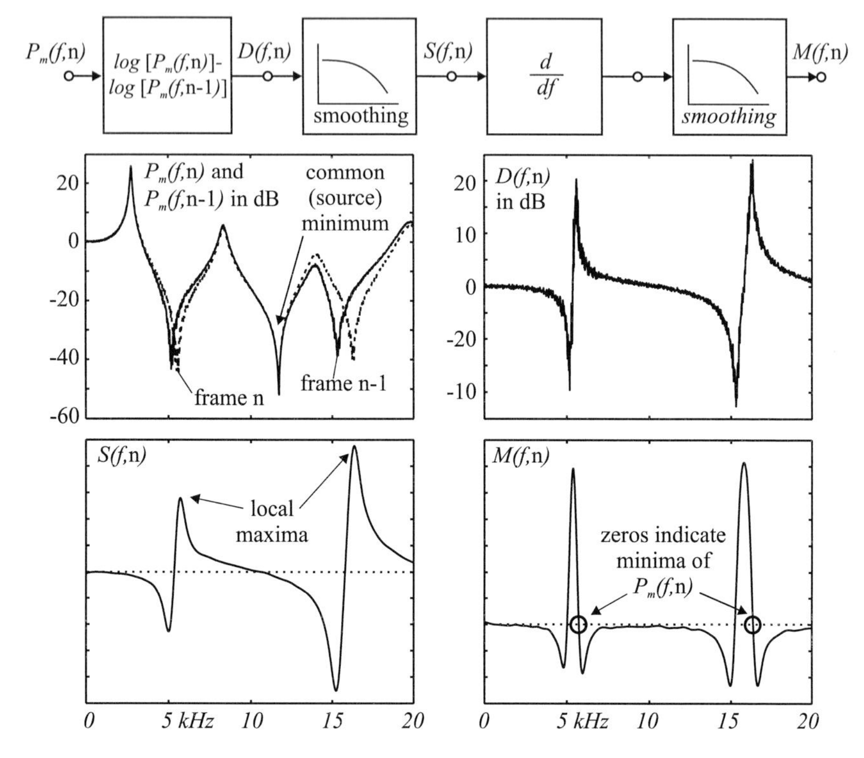

Fig. 42: Algorithm schematic for the detection of the standing wave minima while the probe microphone is inserted into the ear canal (top row). The spectrum panels below show examples to illustrate the algorithm. Top, left: successive pressure measurements $P_m(f,n)$ and $P_m(f,n-1)$ with one common minimum originating from the source and two different standing wave minima. Top, right: level difference spectrum $D(f,n)$; bottom left: smoothed difference spectrum $S(f,n)$; bottom right: minimum detection spectrum $M(f,n)=dS(f,n)/df$. The ordinate axis labels are omitted in the bottom row, because the absolute values of $S(f,n)$ and $M(f,n)$ are not required for minimum detection. The variable n denotes the frame number.

In the microphone spectrum, however, other minima may occur which are present already in the source signal or originate from pinna resonances. Notches in the source spectrum must not be mistaken for standing wave minima, otherwise large estimation errors would occur. The correct minima can be detected efficiently while the microphone is inserted. The frequencies of the standing wave minima vary characteristically depending on the measurement position in the canal.

To determine the minima, variations of the instantaneous magnitude spectra measured at successive time steps have to be detected, while the investigator slowly inserts the microphone into the ear canal. As the minima must be indicated to the investigator while running the experiment, a realtime display of the correct minima is necessary. For this purpose, an appropriate detection algorithm was developed. The structure of the algorithm is shown in Fig. 42. Minima originating from the source are constant for each measurement position, whereas the spectral position of standing wave minima varies over time. Hence, the levels of successive spectral magnitude frames $P_m(f,\mathrm{n})$ and $P_m(f,\mathrm{n}-1)$ are subtracted. The resulting level difference spectrum $D(f,\mathrm{n})$ is an approximation of the time derivative of each magnitude bin in the underlying FFT. Temporal variations in the microphone spectrum change the value of $D(f,\mathrm{n})$ and form characteristical shapes in the difference spectrum. The maximal change occurs at the standing wave minima. The minimum of frame $P_m(f,\mathrm{n}-1)$ is equivalent to the local minimum in $D(f,\mathrm{n})$, whereas its local maximum corresponds to the new position of the minimum in $P_m(f,\mathrm{n})$. The latter has to be detected. The movement direction of the microphone (inward/outward) determines, if the slope at the zero crossing between the two extrema is rising or falling. Unfortunately, variations in the excitation noise signal used in the measurement are enhanced by the differentiation over time as well. Hence, the difference spectrum is smoothed by filtering the spectrum with respect to the frequency axis. To avoid shifting of the spectrum due to the phase response of the applied low-pass filter, it is filtered once forward and once reverse with respect to frequency. Finally, its local maxima are detected by differentiating the smoothed spectrum $S(f,\mathrm{n})$. The zero crossings in the resulting minimum detection spectrum $M(f,\mathrm{n})$ are coincident with the desired minima.

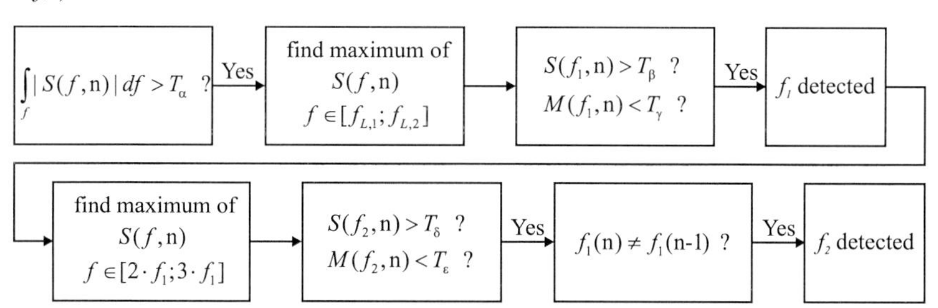

Fig. 43: Schematic of the decision algorithm for minimum frequency detection.

The mentioned algorithm provides fairly good detection results. However, a robust output is obtained only if the microphone is moved, as the first block of the processing chain represents a differentiation over time. The detection process has to be stopped, when the probe is placed at a constant position. Hence, the detection of the local maxima in $S(f,\mathrm{n})$ is controlled using decision stages that enhance the method essentially. The algorithm (Fig. 43) is carried out each time a measurement spectrum frame is processed. The first block decides, whether a microphone movement is present by comparing the total sum over the smoothed difference spectrum with a threshold T_α. If so, the maximum of $S(f,\mathrm{n})$ is detected in the frequency range between $f_{L,1}$ and $f_{L,2}$, which are adjusted to include the expected first minimum frequency. The frequency f_1 is stored if the maximum exceeds the threshold T_β and the absolute minimum detection function is smaller than the upper limit T_γ at the respective frequency. Consequently, the same steps are carried out for the second minimum which is searched only between $2f_1$ and $3f_1$, as it is always located in that range. The frequency f_2 is stored only if the first minimum was altered in the present processing step. It is reasonable to include this necessary condition in the decision algorithm to decrease the impact of high-frequency fluctuations. The thresholds T_α to T_ε are determined empirically.

The instantaneous results of the detection are displayed to the investigator by the developed measurement software. The minima appear as markers on the plot of the current microphone spectrum. Using this method, it is possible to establish a rough estimate of the remaining distance to the tympanic membrane during the measurement ($L = c / 4f_1$). Furthermore, the minima can be placed in sufficiently flat sections of the spectrum which may be important for the damping adjustment (see subsection 3.1.3). As required, the method is not influenced by notches of the source spectrum, because only the standing wave minima vary over time. While the microphone is inserted, a moving average method is used. When the microphone has been placed at an appropriate position, the averaging process is restarted. The new measurement is carried out to determine the microphone pressure signal most accurately (for details on the implemented averaging, refer to Appendix A.4). The minimal frequencies that are used for model fitting are can then be determined precisely from the averaged results. For that purpose, the previously detected minima are used as start values. It turned out that spline interpolation in a small region around the start frequencies provides sufficient results.

After the first two minima have been determined, shape parameters of a sound field model with identical minimal frequencies have to be found. Unfortunately, there is no simple relation between the determined minima and the shape parameters. The chain parameter $\underline{C}_{11}$ of a rigidly terminated cone is given by equation (3.1.3). The irregular curves of the shape parameters r_2/r_1 and L as function of the minimal frequencies f_1 and f_2 shown in Fig. 44 suggest that it is reasonable to solve the inversion problem numerically. In the left panel, the apparent length L is plotted as a function of the first minimal frequency f_1. The second minimal frequency f_2 was used as parameter to obtain an array of curves. The data was calculated using the LMS algorithm described later. The curves are bounded by the two functions $L=c/4\,f_1$ and $L=c/2\,f_1$, which determine the length of a purely cylindrical or conical duct from its first pressure minimum, respectively.

A basic estimation for the length parameter is given by the first-order model $L=c/4f_1$ (represented by the bottom dotted line in Fig. 44, left panel). As approximation function for the parameter r_2/r_1, the function f_2/f_1-2 which represents a linear interpolation between the cylindrical and completely conical case can be applied.

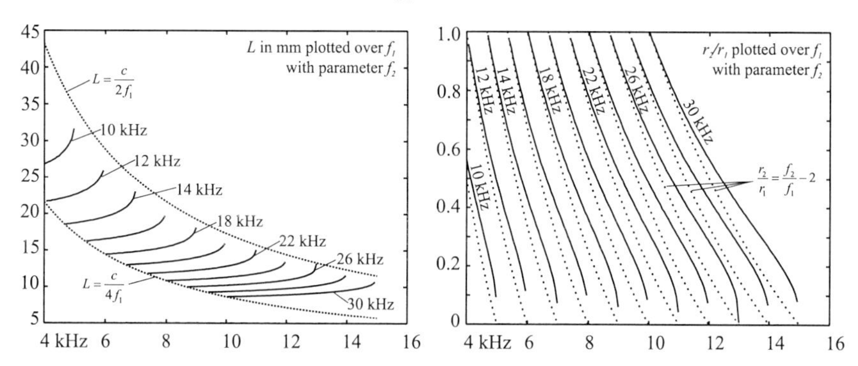

Fig. 44: Shape parameters L and r_2/r_1 as function of the first two minimal frequencies f_1 and f_2.

This estimate is given as dotted set of curves in the right panel of Fig. 44. The figure shows that significant deviations between the real shape parameters and the linear approximations can occur; hence, the correction by numerical procedures is necessary.

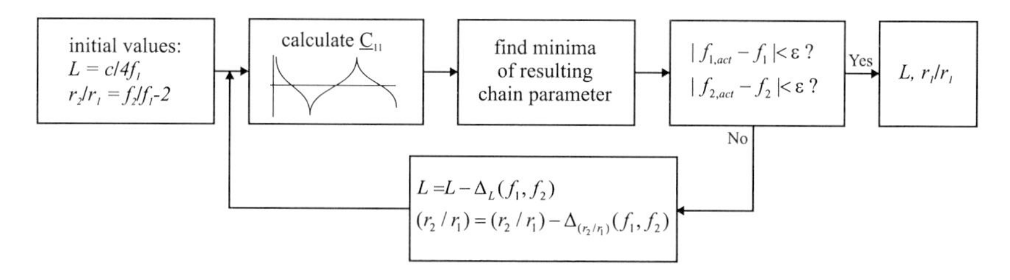

Fig. 45: Schematic of the LMS algorithm used for sound field model fitting

A standard LMS algorithm was applied, which minimizes the mean square error between the required minimal frequencies and the minima of the chain parameter of the resulting model by adapting the values of r_2/r_1 and L iteratively. Fig. 45 illustrates the necessary steps. The initial estimate of the shape parameters is obtained using the simplified expressions described above. The resulting cone is modeled and the first two minima of its input impedance are determined. In a decision step, the minimal frequencies are compared to the measured values. If the difference exceeds a limit, the shape parameters are adapted and the process is reiterated. Fast convergence is obtained, because the fitting function is continous and monotone.

3.1.3 Integration of damping

Although the small amounts of energy that are absorbed by the tympanic membrane are not critical for sound pressure estimation in general (see subsection 3.1.4), the resistive part of the eardrum impedance and the transmission losses of the canal reduce the slope of the minimum both in the transfer function $\underline{H}_{MT}$ and in the measured pressure significantly. As the equivalent system is terminated at first with infinite impedance which does not incorporate losses, the minima are compensated too strong. Hence, the model has to be complemented with resistive elements to reproduce the form of the notches accurately. The presence of the eardrum suggests applying an average eardrum impedance (like the data documented in Hudde and Engel, 1998abc) to terminate the model. The impedance magnitude could then be scaled to compensate the minima adequately. However, even a simpler impedance implementation is feasible. It turned out that a frequency-dependent lumped acoustical resistance R_T terminating the equivalent element is able to model the inherent damping sufficiently well. In the mentioned middle ear impedance, the tympanic membrane and middle ear resonance is modeled. At that frequency, the eardrum impedance is significantly decreased in comparison to higher frequencies. However, the wavelength of sound is sufficiently large to regard the ear canal as lumped element with spatially constant pressure. The resulting measurement error is very small.

In Fig. 46, left panel, the equivalent circuit with termination resistance R_T is shown. The element transfer function (3.1.2) then becomes

$$\underline{H}_{MT} = (\underline{C}_{11} + \underline{C}_{12} / R_T)^{-1} \qquad (3.1.5)$$

with

$$\underline{C}_{12} = \frac{\rho c}{A_2} \frac{r_2}{r_1} j \sin(\beta L) \qquad (3.1.6)$$

The chain parameter $\underline{C}_{12}$ does not exclusively depend on the diameter ratio of the equivalent model, but also on the termination area $A_2=\pi r_2^2$. However, A_2 may be selected arbitrarily because it only represents the reference area of the acoustical termination resistance.

The appropriate resistance value is determined by minimizing the ripples that arise in the estimated eardrum pressure at the minimal frequencies. An example of this process is illustrated in Fig. 46, right panel. The microphone pressure (dash-dotted line) has a distinct minimum at 1.7 kHz. Using the residual transfer function predicted by the lossless equivalent model, the compensation exceeds the minimum (top dotted line with sharp peak). With decreasing termination resistance, the compensation overshoot drops (remaining dotted lines above solid line). At a particular value of damping, the minimum becomes visible again (dotted lines below solid line) and the adjustment direction has to be reversed. The fitting process is finished, when a maximally flat curve occurs (solid line). This correction algorithm can be carried out automatically. Additionally, the phase angle of the signal can be used as fitting criterion, as a characteristic phase step can be observed as long as the system is not adapted correctly.

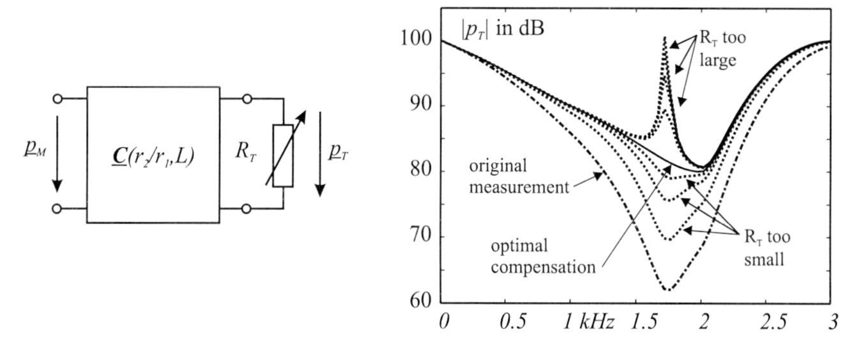

Fig. 46: Left: Equivalent model with adaptation resistance. By variation of R_T, the eardrum pressure frequency response is adjusted to be flat (right panel, solid line). Additionally, the previously measured pressure (dash-dotted line) and different steps that do not compensate the minimum correctly (dotted lines) are shown.

The adaptation resistance R_T is added to the model to cope with the propagation losses of the ear canal and the sound absorption by the tympanic membrane. Strictly spoken, the losses do not only broaden the impedance curve at the minima, but also shift the frequencies of the extrema to some extent. However, this has no meaning for the estimation of the eardrum signal, as the impact of losses is already inherent to the spectrum of the pressure $\underline{p}_M$. The adaptation of the model by adjusting R_T does not alter the minima determined from $\underline{p}_M$.

As already mentioned above, the value of R_T has to be adjusted differently for the each of the two minima. Thus, R_T depends on frequency. The values of R_T between and beyond the two minima are extrapolated to obtain a continuous impedance function. Obviously, the phase angle of the effective impedance is already compensated by the equivalent element $\underline{C}$ representing the residual ear canal sound field. In subsection 3.1.4, it will be shown that the sound field model is able to compensate the shift of minimal frequencies that is caused by complex-valued termination impedances accurately. The frequency-dependent magnitude has to be adjusted separately. This is accomplished by the effective termination resistance R_T. It has to be underlined again that no physical representation of the residual ear canal and its termination is achieved by the equivalent model. Nevertheless, in practice, the resulting values of R_T are generally very large, and the resistance is always larger for the second minimum. This feature can also be found in the magnitude of real eardrum impedances.

3.1.4 Influence of the termination impedance of the ear canal

The described estimation method is based on the assumption that the ear canal is terminated with sufficiently high acoustical impedance. This requirement is generally met by the boundaries near the point **T**, because skin and eardrum tissue are tightly attached to the rigid wall of the bony ear canal (structure-borne vibrations of the skull are negligible in the context of this study). Considering earlier model data (Hudde and Engel, 1998), the tympanic membrane itself can be regarded as high impedance boundary in the hearing frequency range as well. Measurements carried out in the context of that study showed that the impedance of the eardrum is always larger than the approximate duct wave impedance at the same position. Even at the eardrum and middle ear resonance at approximately 1 kHz, the eardrum impedance exceeds the duct wave impedance by 10 dB. Thus, the tympanic membrane guides waves to the end of the ear canal, where a pressure reflection without sign change is obtained for all frequencies (according to the findings of subsection 2.4.2).

However, several studies of the reflection coefficients at the ear canal termination (results are compared in Farmer-Fedor and Rabbitt, 2002 or Feeney and Sanford, 2004) suggest significantly lower impedances than those described in Hudde and Engel, 1998. The pressure reflection coefficient can be calculated as

$$\underline{r} = \frac{\underline{Z}_D - Z_{tw}}{\underline{Z}_D + Z_{tw}} \quad (3.1.7)$$

The power reflectance equals the square of the pressure reflectance ($|r|^2$). The value $Z_{tw}=\rho c/A$ represents the tube wave impedance of the ear canal. The surface A describes the cross-sectional area at the termination position. The termination impedance is denoted as $\underline{Z}_D$, as it is often assigned to the eardrum. If it is considered that the eardrum absorbs energy to a noticeable extent, sound waves are directed towards the eardrum. On the contrary, a rigid tympanic membrane implies that the canal is terminated by the point **T**. Hence, it is not simple to specify the termination position flawlessly. The origin of the differences between the measurements used in Hudde and Engel, 1998abc and the mentioned studies is not certain. Especially frequencies between 1 kHz and 10 kHz are affected. Usually, the first estimation minimum is located in this frequency range. Thus, the influence of the ear canal termination on the eardrum pressure estimation has to be evaluated.

A circuit model of an ear canal according to Fig. 4 was used to determine the minimal eardrum pressure reflection coefficient for which the proposed method is still feasible. The one-dimensional approach was selected, because it is more convenient to adjust the eardrum reflection coefficient in comparison to an FE model. Furthermore, it can be expected that three-dimensional sound field effects have only minor impact on the minimal reflection coefficient. As cross-sectional shape function, one of the ear canal models provided in Stinson and Lawton, 1989 was used. The eardrum impedance was calculated to obtain the desired power or pressure reflection coefficients according to equation (3.1.7). As only the magnitude of the reflection factors was considered in the reference studies, the phase angle of the impedance element is not distinctly specified. Hence, two cases were implemented: (a) a purely resistive impedance which was expected to alter exclusively the slope and the absolute depth of the minima and (b) a complex impedance consisting of a resistance and a mass which influences both the slope and the minimum frequencies significantly. The resulting impedance elements were applied 4 mm anterior to the last port of the model, which is considered to be identical with the point **T** and which was terminated rigidly. The model was excited using constant source pressure and the signal at an arbitrary position **X** was calculated. It was used as virtual "probe microphone pressure" to carry out an estimation of $\underline{p}_T$ using the described algorithm. Finally, the estimated and calculated eardrum signals were compared.

The results are depicted in Fig. 47. In the top row, the influence of the purely real termination impedance is evaluated. A close look at the set of transfer functions between the virtual microphone and the point **T** reveals that not only the slope of the minima is altered by different values of $|r|^2$, but to a very small extent their frequency as well. This effect can be assigned to the applied nonuniform shape function of the ear canal. However, the shift of the minima is not critical, as it is implicitly compensated. The estimation error given in the right panel shows that the deviations are negligible, when $|r|^2$ is set to 0.9 (pressure reflection factor r=0.95). As can be expected, the error increases for smaller values of $|r|^2$. If the power reflection factor at the ear canal termination is adjusted to $|r|^2$=0.3 (r=0.55), the resulting average compensation error is still smaller than 1 dB below 14 kHz and shows a peak of 2 dB below 16 kHz. For this configuration, the logarithmic ratio between the impedance Z_D and the local tube wave impedance equals 10 dB, which represents the minimum reported by Hudde and Engel. The lower power reflectance limit may be allocated at the surprisingly small value of $|r|^2$=0.1 (r=0.32). At this value, estimation errors of 2 dB occur in the whole frequency range. If the reflection factors are reduced further or if negative pressure reflection factors are applied, the estimation method fails.

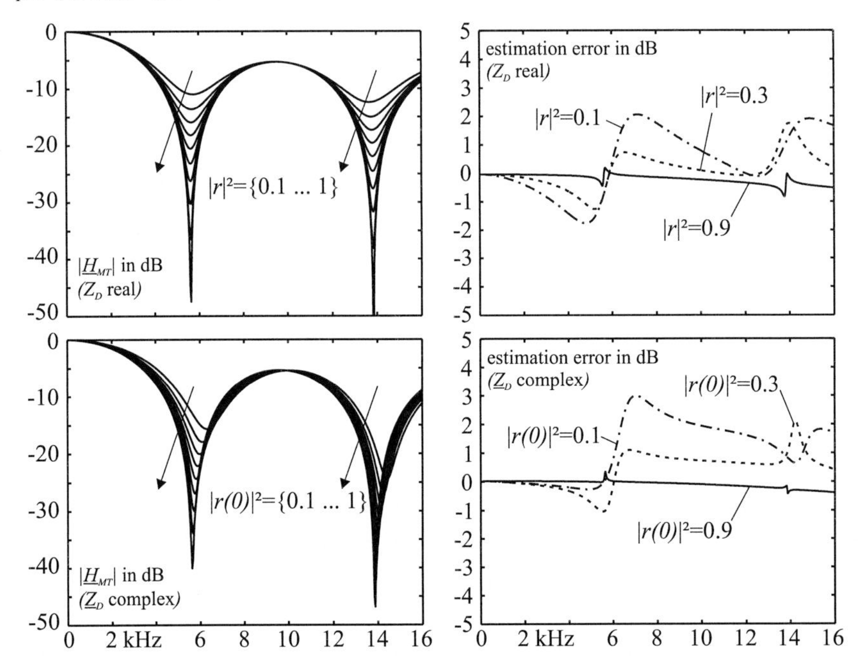

*Fig. 47: Pressure transfer function between the virtual microphone and the point **T** for different magnitudes of the power reflection factor* $|r|^2$ *(left column) and estimation error for the cases* $|r|^2$*=0.9,* $|r|^2$*=0.3 and* $|r|^2$*=0.1 (right column). The arrows in the left column indicate increasing* $|r|^2$*. Top/bottom row: calculation with real/complex eardrum impedance* Z_D*.*

In the bottom row of Fig. 47, the results for the complex impedance are displayed. The complex power reflection factor $|r|^2$ for the frequency f=0 is given as parameter (due to the mass effect, the termination impedance and consequently the power reflection factor increase with frequency). The minima of the transfer function are shifted to higher frequencies with decreasing reflection factor magnitudes. The estimation error is not essentially affected: the functions are similar to those of the real termination impedance case.

The study shows that the estimation method is feasible as long as the incoming pressure wave is reflected at the termination without sign change. This condition is met by the vast majority of individual external ears. The main drop of the reflection factor is found at the eardrum and middle ear resonance. Here, anyway only small mismatches between the pressure signals at the microphone and at the tympanic membrane arise, because the transformation distance is distinctly smaller than the wavelength. As the mentioned measurements consistently show that the reflection factor rises with frequency, the compensation accuracy is significantly increased for higher frequencies.

3.1.5 Excitation of the ear canal sound field

Although the minimum identification method described in subsection 3.1.2 robustly rejects minima that are already present in the excitation spectrum, it is reasonable to avoid distinct notches in the source signal. For the visual adjustment of damping, it is required to distinguish the compensation ripple clearly from the local curvature of the eardrum pressure spectrum, which should be sufficiently flat near the microphone pressure minimum. The distance between the compensated notch and the minimum induced by the source shown in Fig. 46 is already crucially small. Thus, the presence of source minima limits the range of possible estimation frequencies. Furthermore, the compensation of deep excitation notches may result in unnecessarily high amplitudes of the transducer.

The use of loudspeakers as source for the estimation process is feasible exclusively in anechoic environments, as transfer functions in reverberant rooms usually provide a large number of closely adjacent minima. When the speaker is placed very close to the pinna to reduce the influence of wall reflections, other minima may be induced by near field effects. Normally, headphones will be used as sound source for experiments at the external ear. In this case, minima can be generated by duct effects of the pinna (subsection 2.4.1) and by reflections between parallel boundaries in the headphone cup or between the headphone and the tissue. The influence of pinna resonances can be quantified by the sound pressure transfer function between a point close to the membrane and the eardrum as calculated by the FE model of the external ear (Fig. 48).

An inspection of the simulation data shows that the minimum between 9 and 10 kHz is originated by pinna effects. The isosurface structure of the sound field at the minima reveals that the boundary of the *cavum* (*crus helicis*, Fig. 2) represents an acoustical duct, just like the situation discussed in subsection 2.4.1. Obviously, it depends on the incidence direction of sound. The minimum that is found at 14 kHz for the frontal source has the same origin. Sufficiently flat spectra can be achieved by adjusting the headphone position adequately. Due to the broad interindividual variance of pinna shapes and headphones, no general suggestion for cup positioning can be given. If canceling of the minima is by headphone positioning is unattainable, smoothing of the pinna structure using modeling clay should be considered.

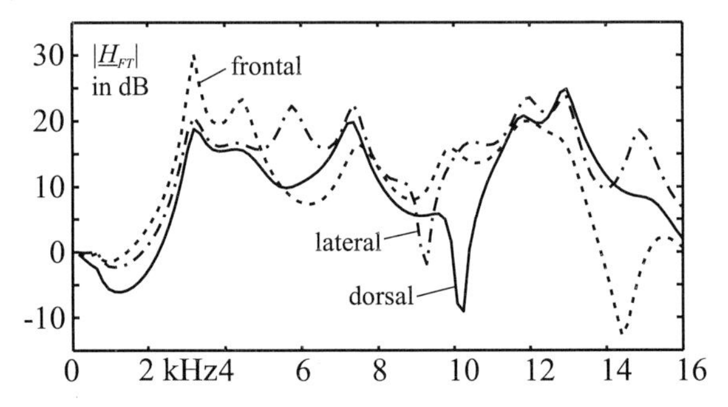

*Fig. 48: Sound pressure transfer function between the respective source membrane (frontal/lateral/dorsal) and the point **T**. The minimum between 9 and 11 kHz depends on the source.*

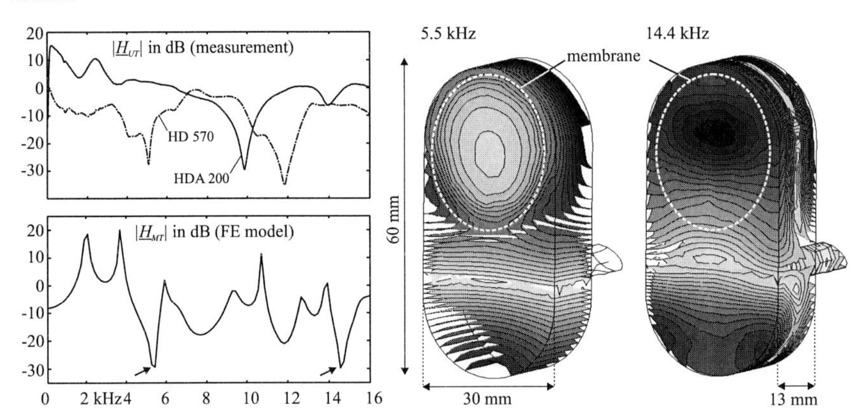

Fig. 49: Measurement of transfer functions $\underline{H}_{UT}$ between headphone voltage and eardrum pressure on an artificial ear (left top panel), pressure transfer function $\underline{H}_{MT}$ between membrane center and eardrum simulation using a simplified FE headphone model (bottom left panel) and sound field structure in the modeled headphone cup for the two minima indicated with arrows.

To analyze the influence of the headphone in more detail, a simplified FE model of a headphone cup with a dummy ear canal was used (Fig. 49, right panels). The panel top left shows transfer functions between the input voltage and the eardrum sound pressure measured with two headphones (Sennheiser® HDA200 and HD570, which represent a closed and an open cup system, respectively). The curves exhibit significant notches. In the transfer function obtained by the FE model, similar minima can be found (left bottom panel), particularly in comparison with the HD570 headphones. The isosurface structure of the sound field shows that the minima originate from standing wave effects in the headphone cup. As the pressure minima correspond to velocity maxima, the reflections between parallel structures in the cup can be avoided by inserting damping foam.

3.1.6 Measurement system

In the following, the measurement system that was used to evaluate the eardrum pressure estimation method is described. The hardware setup was arranged to enable precision sound pressure measurements in the hearing frequency range (a description of the equipment can be found in Appendix A.3). To obtain accurate pressure measurements from positions in the ear canal, specific probe microphones had to be fabricated, which are described in the following. The data acquisition software, programmed with MATLAB®, contains all of the algorithms described above and allows for fast eardrum pressure estimations prior to psychoacoustical measurements. In addition, evaluation tools were implemented.

3.1.6.1 Probe microphones

For measurements in narrow ear canals, probe microphones were assembled. Flexible Teflon tubes of 40 to 70 mm length were attached to Knowles® FG 23629 miniature microphone capsules. The internal and external radius of the tube equals 0.8 mm and 1.2 mm, respectively. The probe tubes represent acoustical ducts that are terminated nearly rigidly at the microphone end; hence strong duct resonances are generated. Thus, the free end was provided with a small amount of foamed damping material. In addition, the use of damping material raises the minima in the acoustical input impedance of the tube. Sound field distortions due to the presence of the probe are reduced. When the probe tip touches the ear canal wall, proximity effects could cause calibration errors. Hence, “domes” as used in open hearing aid fittings were attached to the tips to detach the probe from the canal wall. The schematic assembly of a probe microphone is shown in Fig. 50.

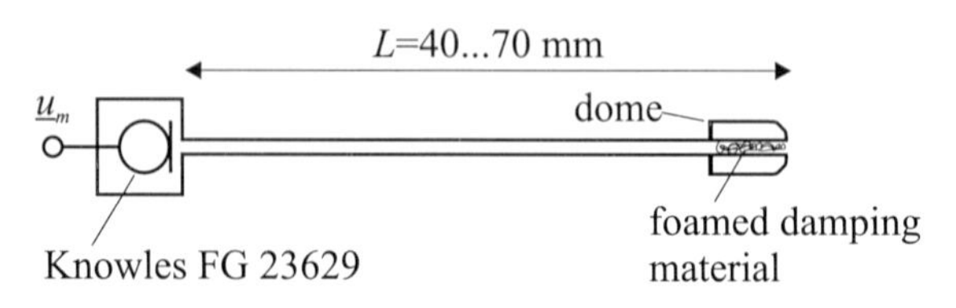

Fig. 50: Schematic assembly of the applied probe microphones

Although the pressure signal is subject to essential damping due to the small inner radius of the tube and the foamed damping material at its orifice, the tube microphones have appropriate signal-to-noise characteristics. Fig. 51 provides corresponding measurement results. In the left panel, the SNR and distortion characteristics are evaluated. The microphones are excited using a sinusoidal signal at 1 kHz using a loudspeaker. The frequency spectrum of the loudspeaker sound pressure is simultaneously monitored using a B&K® 4938 reference microphone. It is secured that no harmonics can be observed in the loudspeaker signal. The miniature microphones obtain a total harmonic distortion (THD) of 0.17 % at full output voltage. The first harmonic appears to be approximately 55 dB below the fundamental. Beyond the second harmonic at 3 kHz, the system noise level is found at approximately -80 dB and below. In conclusion, the miniature microphones have suitable electrical characteristics for the required measurements.

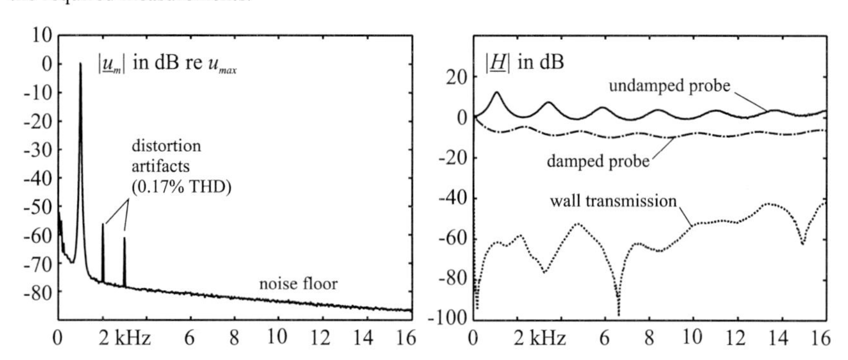

Fig. 51: Left panel: measurements of total harmonic distortion and system noise of an FG 23629 miniature microphone.Right panel: Transfer function between the probe tip and the microphone for damped and undamped probe and wall transmission transfer function for closed probe tip.

In Fig. 51, right panel, different transfer functions of the microphone probes are depicted. The output signal of a probe microphone was related to the output of a miniature microphone without probe. As the microphones were attached to the calibration coupler described in Appendix A.3, both pressures can be regarded as equal. Hence, the results represent the probe transfer function (differences between the used miniature microphones can be neglected for this measurement). As expected, the undamped probe exhibits ripples of approximately 10 dB which can be smoothed by application of foamed damping material. Both functions show surprisingly high transfer factors for high frequencies, which is particularly important for the estimation method. When the probe tip is sealed using putty, the fraction of sound that is transmitted through the elastic tube walls can be examined. For this measurement, the probe tube is coiled up and totally inserted into the coupler. The transfer function shows that a very high damping of external sound pressure is obtained by the tube (-60 dB to -40 dB).

Finally, it was observed that the presence of the damped microphone probe does not alter the coupler pressure more than 0.5 dB in comparison to the sound pressure that arises when the probe bore is filled with putty. As can be expected, the maximum value is located at the duct resonance frequencies of the tube. The measurement results show that the probes are applicable for the required measurements.

For the estimation of $\underline{p}_T$, the probes are inserted into the ear canal and the headphone cups are positioned. It turned out that no further mechanism for the spatial adjustment of the probe tip is necessary. The contact pressure between the skin tissue and the headphone cup is sufficient to fix the probe at a constant position by friction. Nevertheless, the probe can still be moved inward and outward the ear canal by the investigator. When the probes are inserted into the canal, they are bent in an undefined way. However, the FE models of narrow ducts discussed in subsection 2.4.7 show that the probe curvature does not influence the calibration, as the probes have only very small inner diameter.

3.1.6.2 Cross-correlation measurements

For the measurement of pressure signals and transfer functions, a cross-correlation method is applied. The cross-correlation function $R_{xy}(\tau)$ of two signals $x(t)$ and $y(t)$ is specified as

$$R_{xy}(\tau) = E\{x(t) \cdot y(t+\tau)\} \tag{3.1.8}$$

It is a function of the offset τ between the two signals. When $y(t)=x(t)$, the auto-correlation function $R_{xx}(\tau)$ which is a measure of the self-similarity of the signal x results:

$$R_{xx}(\tau) = E\{x(t) \cdot x(t+\tau)\} \tag{3.1.9}$$

By transformation to frequency domain, the cross and auto power spectra $\underline{S}_{xy}(\omega)$ and $\underline{S}_{xx}(\omega)$ are obtained:

$$\underline{S}_{xy}(\omega)=\int_{-\infty}^{\infty} R_{xy}(\tau)e^{-j\omega\tau}d\tau;\ \underline{S}_{xx}(\omega)=\int_{-\infty}^{\infty} R_{xx}(\tau)e^{-j\omega\tau}d\tau \qquad (3.1.10)$$

When $x(t)$ and $y(t)$ denote the input and output signals of a linear and time-invariant system H, its transfer function $\underline{H}(\omega)$ can be obtained by the cross and auto power spectra $\underline{S}_{xy}(\omega)$ and $\underline{S}_{xx}(\omega)$ (Böhme, 1998):

$$\underline{H}(\omega)=\frac{\underline{S}_{xy}(\omega)}{\underline{S}_{xx}(\omega)} \qquad (3.1.11)$$

In the measurement system, an empirical estimation of the cross and auto power spectra and of the transfer function (3.1.11) is implemented by the evaluation of short time spectra with limited frame length N. The short time spectra at the input and output of the system are denoted $\underline{X}(\omega)$ and $\underline{Y}(\omega)$. The empirical estimation of the spectra then holds

$$\hat{\underline{S}}_{XY}=\underline{X}^{*}(\omega)\cdot\underline{Y}(\omega);\ \hat{\underline{S}}_{XX}=\underline{X}^{*}(\omega)\cdot\underline{X}(\omega) \qquad (3.1.12)$$

The data acquisition hardware automatically records N samples from each of the input channels (probe and reference microphones and headphone input signal), triggers a timer event and delivers the digitized sample values to the measurement routine. The data frame is multiplied with a Hanning window function and converted to frequency domain by FFT.

Influences of random noise are reduced by averaging of the empirical approximation. During the minima identification process (subsection 3.1.2), a moving average process is applied which allows for averaging over a limited range of past time. When a suitable probe position is found, a new average process is started which takes all of the input data frames into account. The implementation of the averaging algorithms is documented in Appendix A.4.

3.1.6.3 Calibration

Although they are equipped with damping material at the tip, the probe tubes provide significantly rippled frequency responses (Fig. 51, right panel). To obtain accurate measurements, the measurement system has to be calibrated. For this purpose, the coupler described in Appendix A.3 was used. Fig. 52 shows the transfer functions that have to be taken into account for the estimation process. The B&K® 4938 microphone serves as reference with sensitivity $\gamma=\underline{u}_{mic}/\underline{p}_{mic}$ that is considered to be constant for all frequencies in the regarded range (Fig. 52, left column, top panel). In the coupler calibration measurement (left column, center panel), the transfer function $\underline{H}_C$ between the electrical coupler input signal $\underline{u}_0$ (generated by the measurement system) and the internal sound pressure is determined. It can be calculated from the reference microphone voltage $\underline{u}_{ref}$ by

$$\underline{H}_C = \frac{\underline{u}_{ref}}{\underline{u}_0} \frac{1}{\gamma} \tag{3.1.13}$$

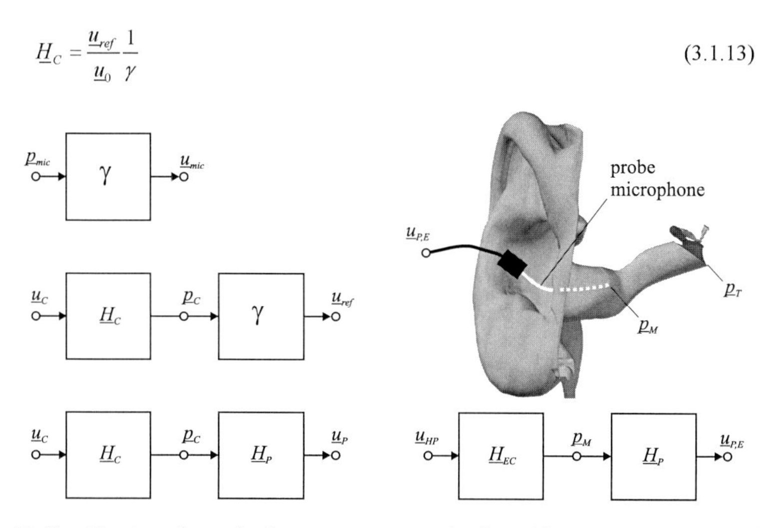

Fig. 52: Specification of transfer functions necessary for the calibration process. Left column, top down: calibration microphone with sensitivity $\gamma=\underline{u}_{mic}/\underline{p}_{mic}$; determination of coupler calibration function $\underline{H}_C$; determination of the probe transfer function $\underline{H}_P$. Right column, bottom: measurement of the probe pressure $\underline{p}_M$

As $\underline{H}_C$ is known, the transfer function $\underline{H}_P$ between the pressure at the probe microphone tip and its output voltage (bottom panel) can be expressed as

$$\underline{H}_P = \frac{\underline{u}_P}{\underline{u}_0} \frac{1}{\underline{H}_C} \tag{3.1.14}$$

Now, the transfer function $\underline{H}_{EC}$ between the headphone input signal $\underline{u}_{HP}$ (generated by the measurement system) and the pressure $\underline{p}_M$ at the measurement position in the ear canal (Fig. 52, right column) can be calculated from the probe microphone output signal $\underline{u}_{P,E}$:

$$\underline{H}_{EC} = \frac{\underline{u}_{P,E}}{\underline{u}_{HP}} \frac{1}{\underline{H}_P} \tag{3.1.15}$$

Using this set of calibration functions, the sound pressure at the tympanic membrane can be adjusted, after the residual ear canal transfer function $\underline{H}_{MT}$ has been estimated according to the described method:

$$\underline{p}_T - \underline{u}_{HP} \cdot \underline{H}_{EC} \cdot \underline{H}_{MT} \tag{3.1.16}$$

3.2 Verification

It is very difficult to verify the estimation method in-situ, because it is not possible to measure the desired eardrum signal directly for reasons that were already discussed in the introduction. Hence, different evaluation approaches were selected that deliver a general view of the quality of the method. Several FE calculations were carried out to study the impact of the microphone probe position and the ear canal shape on the estimation accuracy. To examine the influence of the applied probe tube microphones and to test the practicability of the method, an artificial external ear was fabricated. Furthermore, probe microphone measurements very close to the tympanic membrane were carried out.

3.2.1 Verification using FE models

The preliminary experiments with one-dimensional models of ear canals that were discussed in subsection 3.1.4 show the general feasibility of the estimation method. However, the curvature of natural ear canals and the spatial sound field containing one-sided isosurfaces is not simulated accurately by such models. Precise calculations of pressure values in the canal can be performed exclusively with three-dimensional methods. Thus, the finite-element model of the external ear that was introduced in section 2.3 was used to evaluate the estimation accuracy in various curved ear canals.

This examination also allows for the specification of an optimum insertion depth of the probe microphone. As the discussion of the ear canal sound field in section 2.4 shows, it is necessary that the measurement position is located inside the fundamental field of the canal. The field geometry in this region is independent from the external source by definition; hence, the anterior boundary of the fundamental field can be determined by comparison of the three different near field source types (frontal, lateral and rear excitation). It can be expected that distinct estimation errors occur when a microphone position outside the core region is selected.

Pressure signals from different positions in the FE model of the canal were regarded as probe microphone signal p_M and processed by the estimation algorithm. Identification of the standing wave minima was achieved by inspection of the sound field isosurfaces. The pressure at the termination point **T** was stored as well for comparison with the estimated signals. The probe tube microphone itself was not included in the models. First, the results for the standard ear canal used in the previously discussed FE models are analyzed.

In Fig. 53, the difference between the simulated and the estimated eardrum pressure signals is displayed as function of frequency. The six panels correspond to different first-order model length parameters $L=c/4f_{min,1}$ calculated from the first minimum of the microphone pressure. The parameters are given in the respective plot (in this expression, c=348 m/s was used). It is important to note that the parameter L cannot be interpreted as geometrical distance between two points in the FE model. In comparison with the distance between the measurement position and the point **T** along the approximate middle axis of the model canal, the values of L are generally smaller.

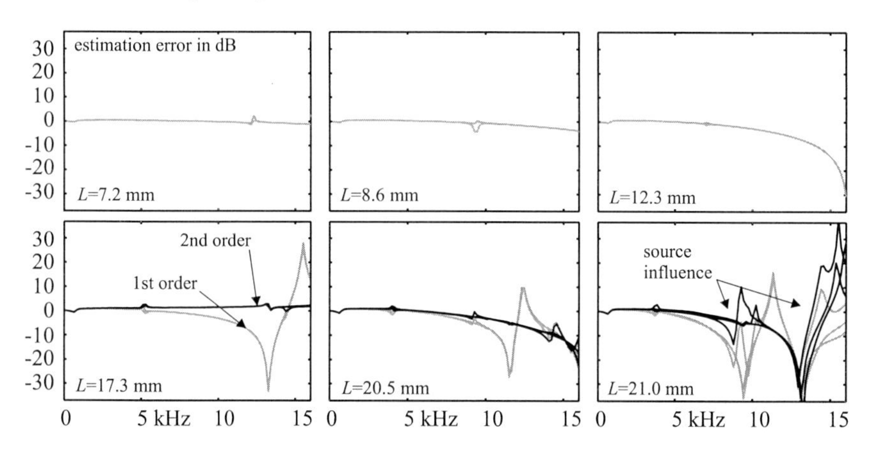

Fig. 53: Errors between estimated and simulated eardrum pressures. Gray lines: first-order model, black lines: second-order model (only in bottom row). The length parameter L was calculated from the first pressure minimum.

The first-order estimations (gray lines) result in a maximum absolute error of approximately 3 dB for the first two transformation lengths. As can be expected, the maximal deviations occur at the first minimum. The error is substantially smaller for frequencies far from the minimum. Between and beyond the minima, the eardrum pressure is almost exactly reproduced. At L=12.3 mm, significant errors can be observed for higher frequencies. The deviation results form the mismatch of the second standing wave minimum frequency in the first-order model prediction. At L=17.3 mm, the second-order method can be applied (black lines), because the second minimum is now located in the simulation frequency range (160 Hz – 16 kHz). The error is comparable to the results of the first-order estimation at L=7.2 mm. The last two cases show strong deviations, as the third and higher minima are not compensated. Furthermore, three different error curves can be distinguished for both models which represent the three applied excitation modes. Obviously, the measurement position lies outside the fundamental ear canal field for these cases, as the source has a strong influence on the results. Thus, the measurements at L=20.5 mm and L=21.0 mm are not suitable for estimation.

Fig. 54 shows first and second order estimations that were carried out at L=7.2 mm and L=17.3 mm, respectively (top and bottom panel in the left row of Fig. 53). In addition, the phase error is displayed (bottom panel). For hearing-related experiments with narrow-band signals, it is often sufficient to adjust only the magnitude of the eardrum sound pressure and to ignore the phase response of the transfer function to the point **T**. Nevertheless, the estimated residual ear canal transfer function compensates the phase change between $\underline{p}_M$ and $\underline{p}_T$ as well. This may be useful when broadband signals are applied (e.g. Schroeder phase signals). The phase error is generally very small. It has maxima at the resonance of tympanic membrane and middle ear and at the probe pressure minima.

It was already stated that the length parameter L is does not represent the geometrical distance between the measurement position and the point **T**. In Fig. 55, top panel, the model parameters are compared to the geometrical distance which was approximated by a spline curve in the FE model.

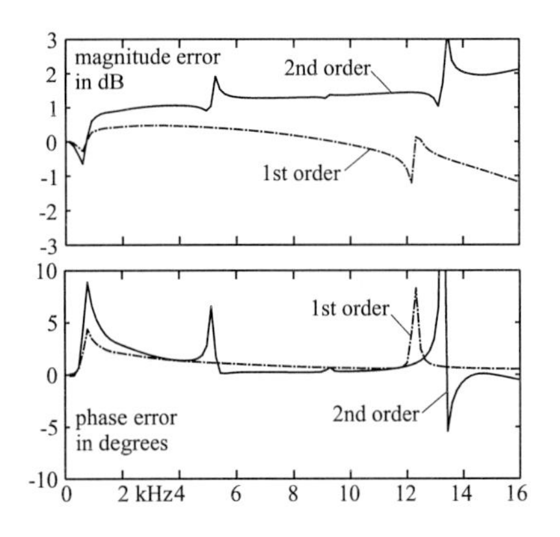

Fig. 54: Estimation error of magnitude (top panel) and phase angle (bottom panel) of $\underline{p}_T$. (first-order estimation at L=7.2 mm, second order estimation at L=17.3 mm.)

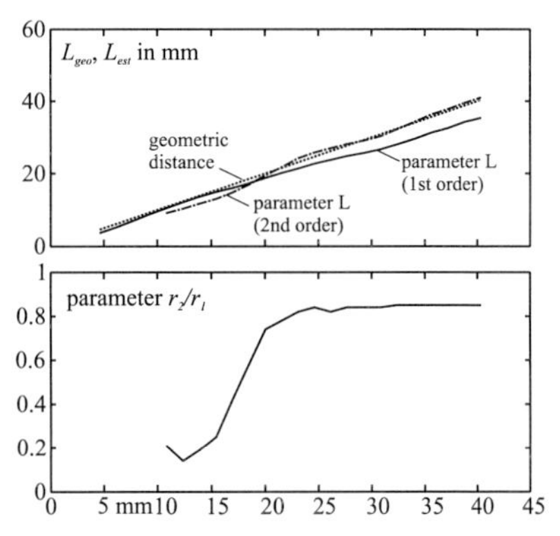

*Fig. 55: Model parameters L (top panel) and r_2/r_1 (bottom panel) as function of the geometric distance between the measurement position and the point **T** in mm.*

The first order model generally underestimates the distance between the microphone and the eardrum. According to Fig. 44, conical deformations of a cylindrical duct increase the frequency of the first minimum. However, the parameter L from the first order model is well-suited to indicate the approximate remaining distance to the tympanic membrane to the investigator while inserting the probe. As the ear canal usually tapers to the point **T**, the true distance is at least as large as indicated by the value of *L*. However, it is not recommended to place the probe tip very close to the eardrum using this process. Good estimations are possible from significantly remote positions, as shown in this section. The length parameter of the second-order model deviates from the geometrical distance as well. At distances between 20 and 30 mm, however, the distance to the tympanic membrane is overestimated. The bottom panel of Fig. 55 shows the variation of the parameter r_2/r_1 with the probe insertion depth. Very close to the eardrum, the residual ear canal appears to be more conical ($r_2 / r_1 \approx 0.2$). The value rises for posterior measurement positions. From a distance of 30 mm on, the parameter remains constant. However, Fig. 53 shows that the respective probe positions are located clearly outside the fundamental sound field.

To analyze the influence of the ear canal shape, the experiment was repeated with various canal models. Due to the broad range of possible geometries, it is difficult to specify parameters for a systematical screening of shapes. Thus, three features that influence the transfer function to the tympanic membrane were selected: (a) the canal length (long, medium, short), (b) the curvature (straight, curved) and (c) the inclination angle of the eardrum (flat, steep). By combining the features, 12 different geometries were created manually. The ear canal models and the respective estimation error plots can be found in Appendix A.5. Up to 12 kHz, the absolute error is below 2 dB for each of the canal models. For some cases, the deviation is significantly smaller. The error increases with frequency, because higher order minima are not compensated correctly. Peaks of error can be found at the estimation minima. The first-order and second-order models generally yield different errors. However, the estimations are based on different virtual microphone positions. The comparison of the plots suggests that the accuracy of the estimation results does not depend significantly on the shape of the ear canal.

3.2.2 Verification using measurements in an artificial ear

In reality, the presence of a microphone or probe tube affects the ear canal sound field. Such effects are not modeled in the finite element simulation. To examine the influence of the probe and to test the practicability of the method, measurements with an artificial ear were carried out. The model consists of a silicone pinna replica molded from a natural ear (see Fig.5). The ear canal was first formed using putty regarding the shape of the FE ear canal model.

After that, it was cast with synthetic resin and the putty was removed. The canal is terminated rigidly by an acrylic glass plate with a bore shaped like an eardrum. In the physical model, $\underline{p}_T$ is measured directly by an additional reference probe that is placed at the point **T**. The sound field in the model is excited by damped headphones. From different microphone positions, estimations were carried out, starting at the innermost point and pulling the microphone successively out of the ear canal.

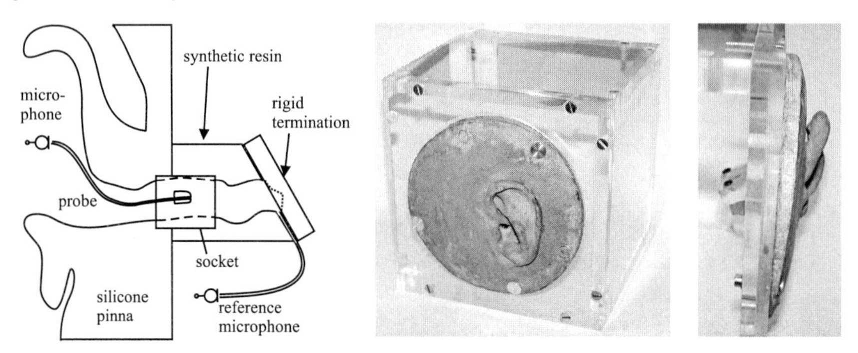

Fig. 56: Artificial ear for verification measurements, left panel: schematic design, center panel: artificial ear in acrylic box, right panel: detail view on the ear canal block.

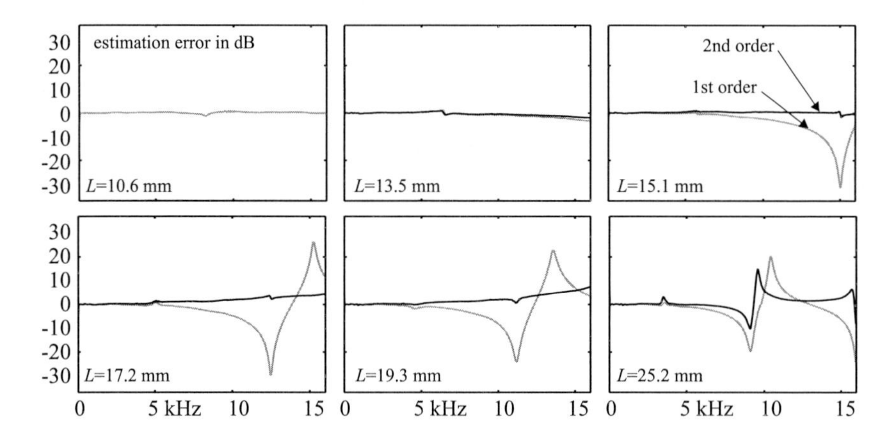

Fig. 57: Errors between the estimated and the measured eardrum pressures. Gray lines: first-order model, black lines: second-order model. The length parameter L was calculated from the first pressure minimum.

Fig. 57 displays the difference between the estimated eardrum sound pressure and the direct measurement for six microphone positions. In each panel, the corresponding length parameter of the first-order model is given. The errors are closely related to those found in the finite-element simulation.

It turns out that it is possible to determine the eardrum pressure within error limits of 2 dB, when the appropriate model is chosen. A second-order estimation is possible at L=13.5 mm already, because an extended frequency range was used in the measurement (0 – 22 kHz). Thus, the second minimum can be analyzed for shorter estimation distances than in subsection 3.2.1. The results show that the presence of the microphone does not affect the estimation process critically.

Furthermore, the setup can be used to examine how the probe presence in the ear canal influences the sound pressure at the tympanic membrane. In Fig. 58, the variation of the eardrum pressure $\underline{p}_n$ for different probe positions in the ear canal is depicted. The signals are related to the eardrum pressure $\underline{p}_0$ arising when the probe is placed at the most anterior position. Obviously, the probe placement affects the eardrum pressure with several peaks and notches of approximately 5 dB. Of course, the effect is compensated implicitly by the pressure estimation method. The results show, however, that uncompensated measurements are influenced significantly by the presence of a probe.

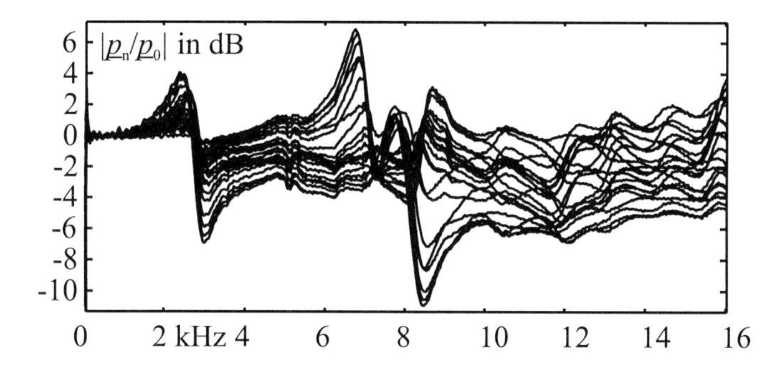

Fig. 58: Variation of the eardrum pressure for different probe microphone positions in the ear canal.

3.2.3 Discussion of the estimation error

To compare the effects of the excitation source, the probe insertion depth and the selection of a first- or second-order model, the absolute error was averaged over the frequency range up to 16 kHz. In contrast to Fig. 53 and Fig. 57, it is displayed over the geometrical distance between the measurement location and the point **T** (Fig. 59). In the finite-element model (top panel), this parameter was determined by scaling the length data by the curve shown in Fig. 55. For the artificial ear measurement (bottom panel), it was estimated by the distances at which the microphone was successively pulled out of the ear model. Due to the unequal frequency ranges, the curve sections of the simulation and the measurement cover different distance ranges. The second-order model was applicable much earlier with the artificial ear, because the upper frequency limit was significantly higher than in the FE simulation.

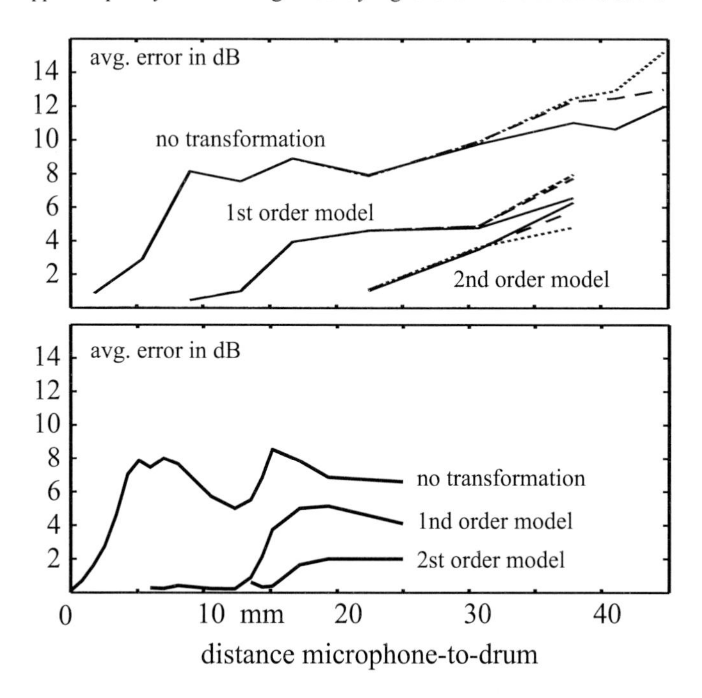

*Fig. 59: Average error between the estimated and the simulated or measured eardrum pressures as a function of the distance between the microphone and the point **T**. Top panel: finite-element model, the three excitation cases are visualized by different line styles. Bottom panel: measurement in an artificial ear.*

In the top panel, the three excitation types are indicated by different line styles. The core region begins, where the influence of the excitation direction becomes visible, which is found at a distance of approximately 30 mm.

If the microphone pressure is taken as identical to the eardrum pressure and thus no transformation is carried out, a distinct error occurs even at very short distances to the tympanic membrane. Of course, basically high frequencies are affected. The simple first-order correction yields significantly better results. However, it does not compensate the second minimum correctly, thus the error increases rapidly when the second minimum is located in the regarded frequency range (at approximately 12 mm).

As expected, the second order estimation reduces the error essentially. A tip-to-eardrum distance of approximately 15 mm seems sufficient to provide very good accuracy. This corresponds to a first minimal frequency between 5 and 10 kHz, depending on the tapering of the ear canal. In conclusion, some basic rules for the optimal minima frequencies can be given that can be considered when the probe is inserted:

- The first probe pressure minimum should be located between 5 kHz and 10 kHz.
- The second minimum should be placed inside the measurement frequency range (otherwise, a first-order estimation can be carried out).
- The eardrum pressure should be sufficiently flat near both minima to enable a precise adjustment of the damping.

3.2.4 In-vivo measurements in the ear canal

Neither the FE simulation nor the artificial ear measurement provides perfectly realistic representations of the ear canal sound field. However, it is impossible to carry out in-vivo measurements of $\underline{p}_T$ directly at the eardrum. To approximate such in-vivo experiments, the estimation method was evaluated by measurements at positions not directly at the point **T**, but very close to the tympanic membrane.

3.2.4.1 Transfer functions in the canal

In this experiment, estimations of the ear canal transfer function in the eardrum coupling region are compared. A reference probe was inserted deeply into the ear canal until the subject reported eardrum contact. The probe was then retracted by a small amount. In addition, an estimation probe was inserted. Two second-order eardrum pressure estimations were carried out at different probe tip positions. After that, the transfer functions between the respective reference pressure measurements in close vicinity to the tympanic membrane and the estimated eardrum signals were calculated. The results are shown in Fig. 60. In the top row, both estimations are displayed for three measurements. The results obviously represent the transfer function of a duct terminated with very high acoustical impedance.

In the bottom row, the logarithmic ratio of the two transfer functions is depicted. The curves have common features with the error calculations presented in the previous sections: the estimation results are very similar, except at the estimation minima, where error peaks of approximately 2 dB magnitude arise (in each experiment, the first estimation minima are located in the frequency range from 4 to 8 kHz).

Although the absolute difference does not exceed 2 dB in the frequency range from 0 to 12 kHz, the accuracy seems to deteriorate above 8 kHz. To some extent, the deviations can be traced back to small shifts of the reference probe tip position due to the movement of the estimation probe.

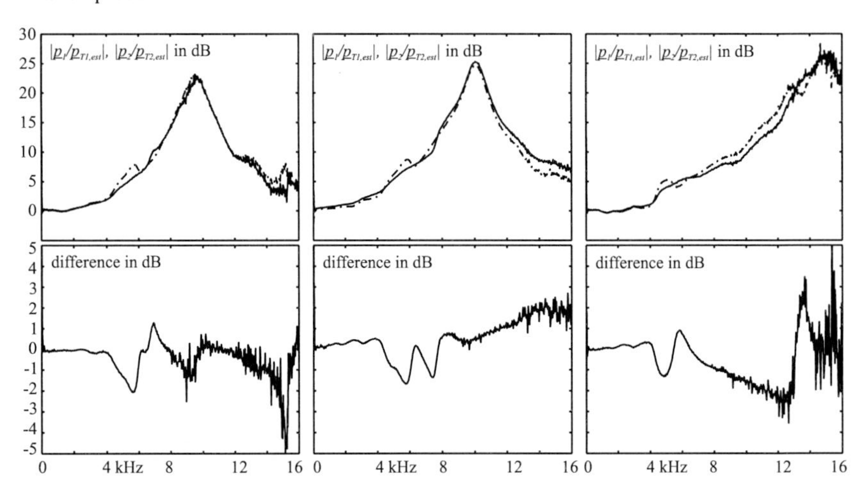

Fig. 60: Comparison of estimated eardrum pressure with pressure measurements at three positions close to the eardrum (top row). For each measurement, two estimations with different probe insertion depths were carried out (represented by different line styles). Bottom row: difference between the two estimations.

3.2.4.2 Detection of one-sided isosurfaces

Evidence of significant spatial sound field distortions (for example one-sided isosurfaces) can be obtained easily using two probe microphones that measure the sound pressure at closely adjacent points. The measured signals are identical, if the probe tips are located on the same isosurface of the sound field. When frequency-dependent one-sided isosurfaces occur, deviations can be expected.

Two microphone probes were attached to each other in such a way that the tips were immediately adjacent (1.2 mm distance between the centers of the probe cross-section). The ratio between the two probe signals was then determined in the sound field inside of the calibration coupler (see Appendix A.3). This function was used for calibration. The double probe was inserted into the ear canal of a human subject. At several positions, the ratio between the two probe signals was measured again and related to the previously recorded calibration function. The external sound field was excited by two switchable miniature loudspeakers. To implement two significantly different excitation fields, one speaker was placed below and the other in front of the pinna. Fig. 61 shows four characteristic results for the pressure magnitude ratio between the two probe pressure signals. The two sources are indicated by different line styles.

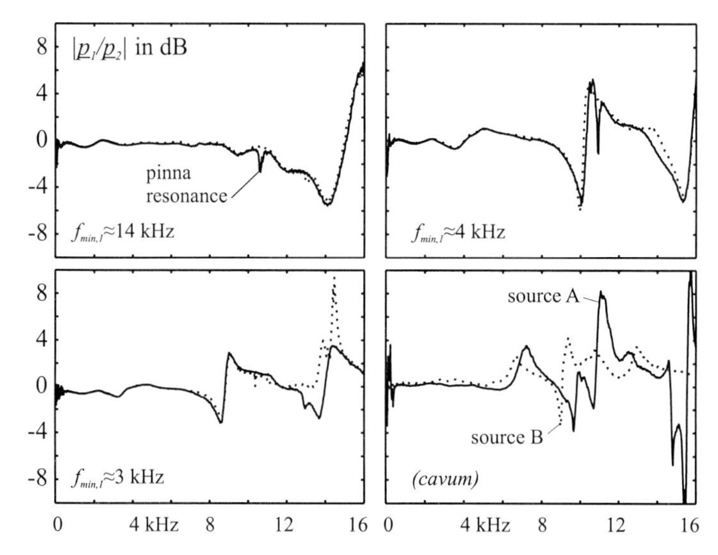

Fig. 61: Measurements of two directly adjacent pressure signals at four different positions in the canal indicated by $f_{min,1}$. The two line styles correspond to the two applied sources. At approx. 10 kHz, no accurate measurement is possible due to resonances in the pinna which essentially reduce the pressure in the canal (note the notches in the top row panel).

In the panels, the approximate frequency of the probe pressure minima is given. Due to the presence of one-sided isosurfaces, the minima do not occur exactly at the same frequency. Even close to the eardrum, significant deviations between the probe signals can be observed (top left panel). Such deviations can be found at other positions in the canal as well. As expected, the deviations at the standing wave pressure minima increase with frequency. The sharp notch that occurs at 9 kHz in the two panels is most likely assigned to a pinna resonance that is excited by the source below the pinna only. In the very narrow frequency range, the excitation signal of the probes has very low energy. Hence, the measurement accuracy deteriorates, which was also indicated by the coherence function of the signals (not shown here). Furthermore, the panel indicates that the isosurfaces depend on the source selection (note the differences above 12 kHz). Hence, the probe is not longer located in the core region. For a position in the *cavum conchae* that is clearly outside the ear canal (bottom right panel), the probe signals are distinctly different.

The results of this experiment suggest that the double-probe measurement can be used to identify the entrance of the core region in individual measurements, if required.

3.3 Measurement of equal-loudness level contours with eardrum pressure reference

3.3.1 Equal-loudness level contours

Perceived loudness is a function of frequency and of the sound pressure level (SPL) at the ear. Equal-loudness level contours are curves in a pressure-over-frequency diagram that represent the same loudness level. The unit of loudness level is the “phon”. The contours can be determined by measuring points of subjective equality (PSE) of variable test sounds and constant reference signals. Subjects have to compare test stimuli to an anchor stimulus. In standard experiments concerning loudness curves, the anchor stimulus is represented by a sinusoidal tone with a frequency of 1 kHz. The subjects control the sound pressure level of the test tone in such a way that the perceived loudness of both tones is equal. By variation of the test tone frequency, frequency-dependent curves can be determined.

Measurement conditions and requirements are standardized (e.g. ISO226). It is common practice to specify the free field sound pressure that arises at the position of the subject’s head (while the subject is absent) as a reference signal for loudness PSE measurements. In this configuration, the pressure spectrum at the eardrum shows significant peaks and notches due to the transmission characteristics of the ear canal. In addition, further external ear effects like pinna resonances or reflection and diffraction at head and torso become effective. Equal-loudness level contours yield significant notches at the resonance frequency of the ear canal. The exact frequency depends on the canal length and geometry, which is interindividually different. To obtain a set of standard curves like those used for the ISO226, it is necessary to average the contours determined for a large number of subjects. This operation blurs and broadens the individual minima (see Fig. 62). The first minimum is located at 3-4 kHz, the second minimum can only be recognized rudimentarily as decline of the curve at 12 kHz.

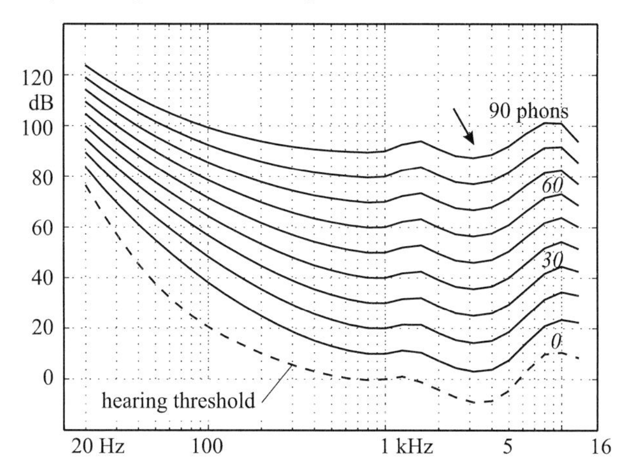

Fig. 62: Equal-loudness level contours as standardized in ISO 226. At 1 kHz, loudness and pressure level are identical. The bottom contour represents the hearing threshold. The arrow indicates the broad minimum that originates from averaging over a large number of subjects.

When the eardrum pressure is applied as a reference signal, minima and maxima of the external ear transfer function are compensated. The resulting equal-loudness level contours can be expected to be free of the ear canal transfer function ripples that occur when the curves are recorded using free-field sound pressure as reference. Consequently, averaging over contours related to the eardrum pressure is essentially more meaningful. To analyze this, monaural equal-loudness level contours with relation to the sound pressure at the tympanic membrane were determined.

3.3.2 Determination of discrimination thresholds

To achieve a robust and efficient estimation of the loudness thresholds, a two-alternative forced-choice (2AFC) procedure was applied (Green, 1993; Dai, 1995; Ulrich and Miller, 2004). In 2AFC experiments, the subject has only two response options related to the considered stimulus feature. For the determination of discrimination thresholds, both anchor and test stimulus have to be presented. The subject judges the perception by answering a simple question (e.g. "Which stimulus was louder?"). It is allowed to replay the stimuli, but it is not possible to answer indifferently (forced-choice method). After the response is given, the measurement system adjusts the physical features of the test signal to reduce the perceived difference between the anchor and the test stimulus (in this context, the sound pressure level is adjusted). In the following, the implemented procedure is described in detail. Similar setups were used in many previous investigations (e.g. Fastl et al., 1990; Gabriel et al., 1997; Reckhardt and Mellert, 1998; Reckhardt, 2000; Takeshima et al., 2001).

Test and anchor stimulus are always presented in pairs. To cancel any influence of the stimulus order, the sequence of the signals is randomized. At first, the test stimulus is presented with an initial pressure level in conjunction with the anchor stimulus at 1 kHz that features the desired loudness (at 1 kHz, sound pressure level and loudness level are equivalent per definition). If the test stimulus was perceived louder than the anchor stimulus, its sound pressure level in the next experiment is reduced by an initial step size. After a number of repetitions, the loudness of the test tone falls below the loudness of the anchor stimulus. Now, the subject will necessarily reverse the direction of SPL adjustment. The trace of applied SPL values obtains a turning point. At each turning point, the step size of the adaptation is reduced by half. The cycle is repeated iteratively, until a final condition (minimal step size, limit of turning points etc.) is reached. In the course of an experiment, a path connecting the turning points results, which approaches the ideal PSE.

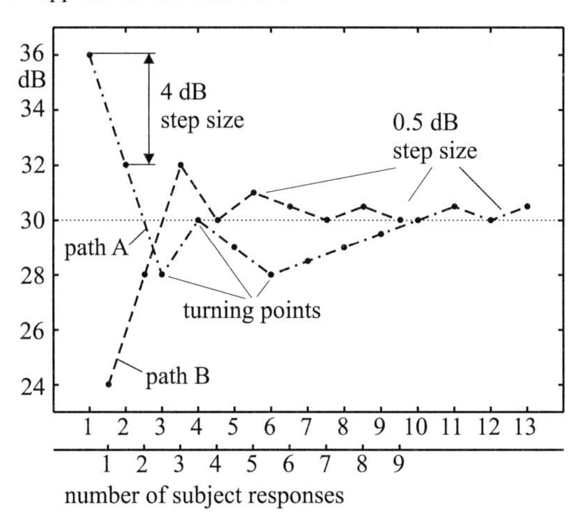

Fig. 63: Example of a two-path 2AFC experiment. The two paths are indicated by different line styles. Each path shows the variation of the test tone sound pressure level. In the example, the expected threshold is located at 30 dB. The measurements are shown in alternating order for clarity, although responses of the two paths are randomized in real experiments.

It is well known that the PSE estimations for loudness discrimination obtained by the 2AFC method depend systematically on the selected initial value (Gabriel et al., 1997). It is particularly important, whether its loudness is above or below the examined threshold. Thus, two paths are recorded simultaneously with alternating response requests. The initial level of one path significantly falls below the expected threshold, while the initial level of the other path exceeds it. The order of the path presentations is randomized. Thus, conscious influence of the subject is inhibited as far as possible.

To illustrate the method, two paths of an example experiment are displayed in Fig. 63. An initial step size of 4 dB was applied. The paths A and B are initialized with stimulus levels of +6 dB and -6 dB with respect to the assumed threshold (30 dB). In the example, the expected threshold is identical to the real threshold. Both paths exhibit a turning point at the third subject response. At the turning points, the step size in dB is divided by two. When a step size of 0.5 dB is achieved, it is not decreased further. Each path is stopped after 6 turning points. While the first path requires 13 responses, the second path is already stopped after 9 responses. The resulting threshold can be calculated by different measures. In this study, the average of the turning points in the region of the finest step size was used. For the example in Fig. 63, this method yields a threshold of 30.6 dB.

As already mentioned, the step size of the paths does not decline below 0.5 dB. A higher resolution is not reasonable, because this value approximately represents the just noticeable difference (JND) of loudness at 1 kHz and 30 dB SPL (Zwicker and Fastl, 1999).

3.3.3 Measurement setup

The initial sound pressure of the test stimuli was chosen identical to the respective anchor loudness (e.g. 30 dB SPL at the eardrum for the 30 phon contour). Thus, the two paths start at 24 dB and 36 dB. As the equal-loudness level contours depend on frequency, it can be expected that the ideal initial values should also be increased for very high and low frequencies. The range of very low frequencies (20 Hz to 250 Hz) that is usually included in equal-loudness level studies is omitted to reduce experimental effort. It can be expected that the sound pressure around the head and in the ear canal is constant for these frequencies, thus eardrum reference does not alter the results. To test the feasibility of the estimation method for very low frequencies, a separate experiment was carried out (subsection 3.3.4.2). Hence, only the high frequency increase was implemented.

As stimuli, sinusoidal signals of 1 s duration are presented. The stimuli are switched on and off by hanning window slopes with 50 ms rise/fall time. The pause between the stimuli is 1000 ms. To control the sound pressure of anchor and test stimuli, an eardrum pressure calibration according to equation (3.1.16) was carried out before each measurement. The stimuli were weighted according to the determined transfer function. The anchor stimulus had a frequency of 1 kHz in all experiments.

The measurement software was integrated into the framework for eardrum sound pressure estimation. Thus, contours of equal-loudness level could be recorded immediately after an ear canal calibration was performed. The user interface consisted of three buttons ("Sound 1 is louder", "Sound 2 is louder", "Repeat", see Fig. 64).

In synchronization with the successive presentation of the two stimuli, the color of the respective button was changed to green. The additional indication is necessary, because the loudness of test stimuli may be below the absolute hearing threshold. Thus, only one sound would be perceived. To avoid the click noise of the computer mouse button, keyboard control was implemented. Upon data entry, the color of the selected button was temporarily changed to red for feedback to the subject. In the text field entitled "Progress", the consecutive number of the current sub-experiment (i.e. 2AFC determination of the loudness PSE at one frequency) was displayed. After each sub-experiment, a white noise signal with hanning amplitude envelope was presented to the subject (1 s duration, approx. 40 dB sound pressure level). This neutral stimulus was delivered to indicate the termination of the current sub-experiment and to focus the attention of the subject to the new pass.

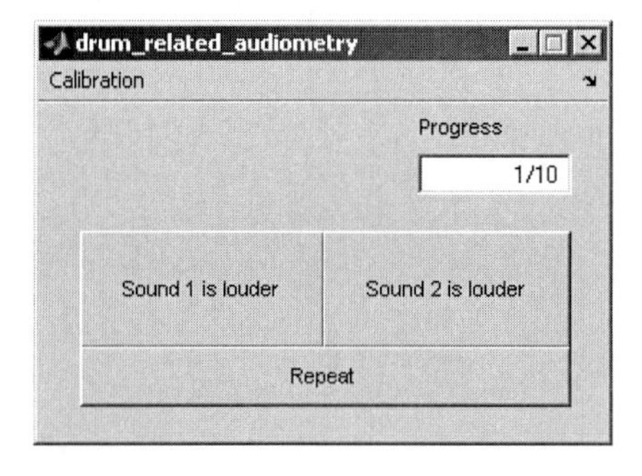

Fig. 64: User interface for the 2AFC test

All experiments were carried out in a sound-proof cabinet. Damped, open headphones (Sennheiser® HD570) were used monaurally; the contralateral ear canal was occluded using an earplug to avoid cross-talk effects. The probe microphone used for pressure estimation was left in the ear canal during the psychoacoustical measurement. The subjects were instructed to maintain a constant position of the headphone. Prior to the experiments, it was asserted that the subjects had no significant hearing loss in the regarded frequency range using a commercial audiometer.

3.3.4 Experiments and Results

3.3.4.1 Influence of the probe insertion depth

In the pilot experiment, the influence of different microphone positions on the results was evaluated. Three series of measurements were carried out with one listener (subject 1, aged 30). In the first series, the frequency of the first minimum was chosen near 5500 Hz (±200 Hz). For the second series, the probe was inserted deeper, yielding a first minimal frequency of 7500 Hz (±200 Hz). In both cases, the eardrum sound pressure was estimated using a second-order model. In the third series of measurements, the probe tip was brought into close vicinity of the tympanic membrane (first pressure minimum at approximately 20 kHz), and a first-order model was applied.

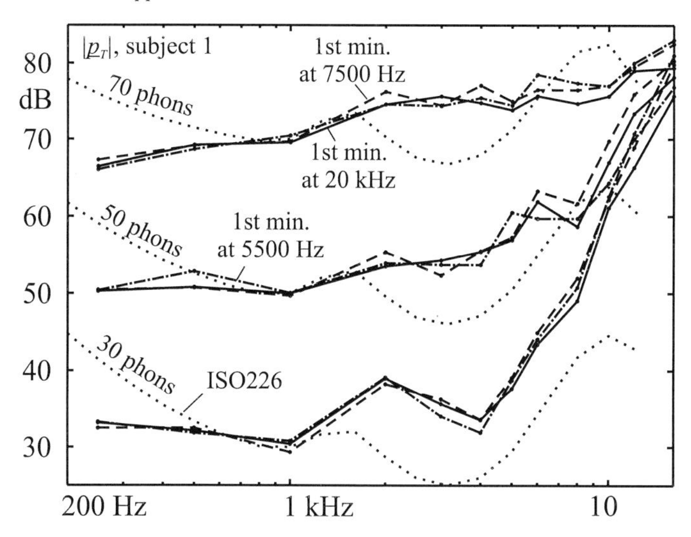

Fig. 65: Equal-loudness level contours at 30, 50 and 70 phons with reference to p_T for three microphone positions.

Points of equal loudness perception were recorded at 30, 50 and 70 phons for a set of 12 frequencies (250, 500 Hz, 1, 2, 3, 4, 5, 6, 8, 10, 12, 14 kHz). The frequencies were presented in random order. For each microphone position and anchor loudness, four measurements were averaged. Thus, a total number of 36 single experiments results. Additionally, 2-4 training passes were carried out prior to the averaged measurements. To avoid fatigue, only two measurements were performed per day. Between the passes, sufficiently long rest periods were maintained. The determined equal-loudness level contours are shown in Fig. 65, along with contours according to the ISO226 standard for comparison.

For low frequencies, the measurements have a surprisingly large deviation from the ISO standard (approximately 10 dB). As shown later (subsection 3.3.4.2), the deviation is assigned to the design of the experiment. At 1 kHz, the measured loudness level is per definition nearly identical to the sound pressure level. As expected, the curves determined with reference to the eardrum sound pressure are free of the ear canal minimum between 2 and 4 kHz. The contour for 30 phons shows a significant peak at 2 kHz. This feature is examined with higher frequency resolution in subsection 3.3.4.4. Regarding the probe tip position, the contours show only minor deviations (max. 3 dB below 10 kHz). These error bounds could be confirmed with two further subjects (Fig. 66).

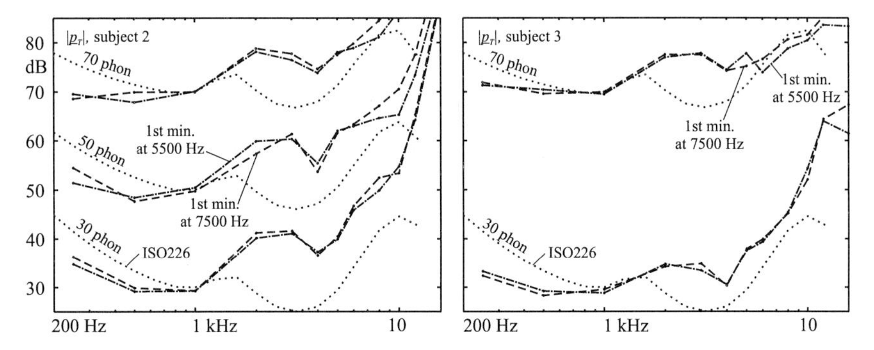

Fig. 66: Equal-loudness level contours at 30, 50 and 70 phons with reference to p_T estimated from the two remote microphone positions according to Fig. 65.

The experimental setup was identical to the previous listening test. However, the estimation pressure measurement close to the tympanic membrane was omitted for the safety of the subjects. Subject 3 only recorded equal-loudness level contours for 30 and 70 phons. In the resulting data, the deviation at low frequencies can be found for both subjects. Again, the ear canal minimum between 2 and 4 kHz is cancelled. The difference between equal-loudness level contours basing on estimation signals from different positions in the ear canal is very small (max. 3 dB below 9 kHz, max. 6 dB below 16 kHz). The standard deviation of all measurements (subjects 1,2,3) was below 3 dB, except for the 50 phons contour of subject 2 (first minimum at 7500 Hz), which was within limits of 6 dB for frequencies above 6 kHz. Standard deviation bars are not displayed in the figures for clarity.

3.3.4.2 Low-frequency measurement and comparison with free-field conditions

The curves that were recorded in the preliminary experiments show large deviations from the standard contours at low frequencies. However, values that are similar to the determined low-frequency contours were published in other studies as well (Fastl et al., 1990). It is well known that the design of experiments for equal-loudness level contours may influence the recorded perception thresholds essentially (Gabriel et al., 1997). In particular, the range of test tones that occur during a measurement of a particular data point has significant impact. To investigate whether the deviations are caused by the estimation method or by range effects, an experiment was performed at low frequencies (63, 125, 250 and 500 Hz) with one listener (subject 1). An anchor loudness of 30 phons was selected, because the deviations are most significant at low loudness levels. The contour was recorded under three different conditions: (a) measurement with relation to the estimated eardrum pressure, 30 dB initial level for all frequencies, (b) measurement with relation to the estimated eardrum pressure, initial level adapted according to the standard contours, (c) free field measurement in an anechoic chamber according to the specifications of ISO 226, 30 dB initial level for all frequencies. After 2-4 training passes were carried out, 4 measurements were performed for each condition. The averaged results of the experiment are depicted in Fig. 67.

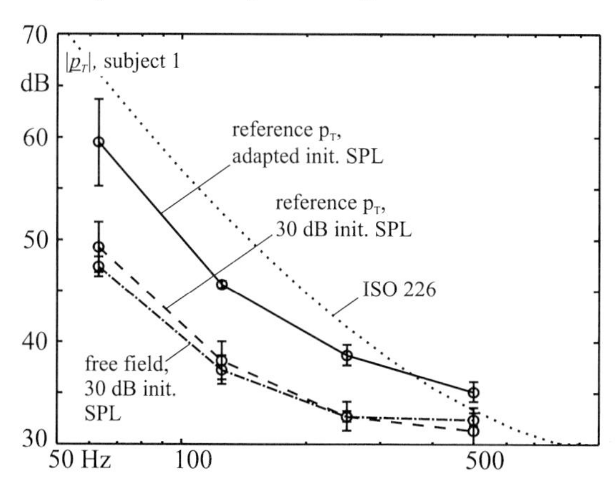

Fig. 67: Equal-loudness level contours at 30 phons for low frequencies (subject 1). The vertical bars denote the standard deviation.

The contours determined with an initial sound pressure level of 30 dB have only very small difference. As expected, the low frequency results do not depend on the reference signal. The contour determined with adapted initial SPL is located at significantly higher values

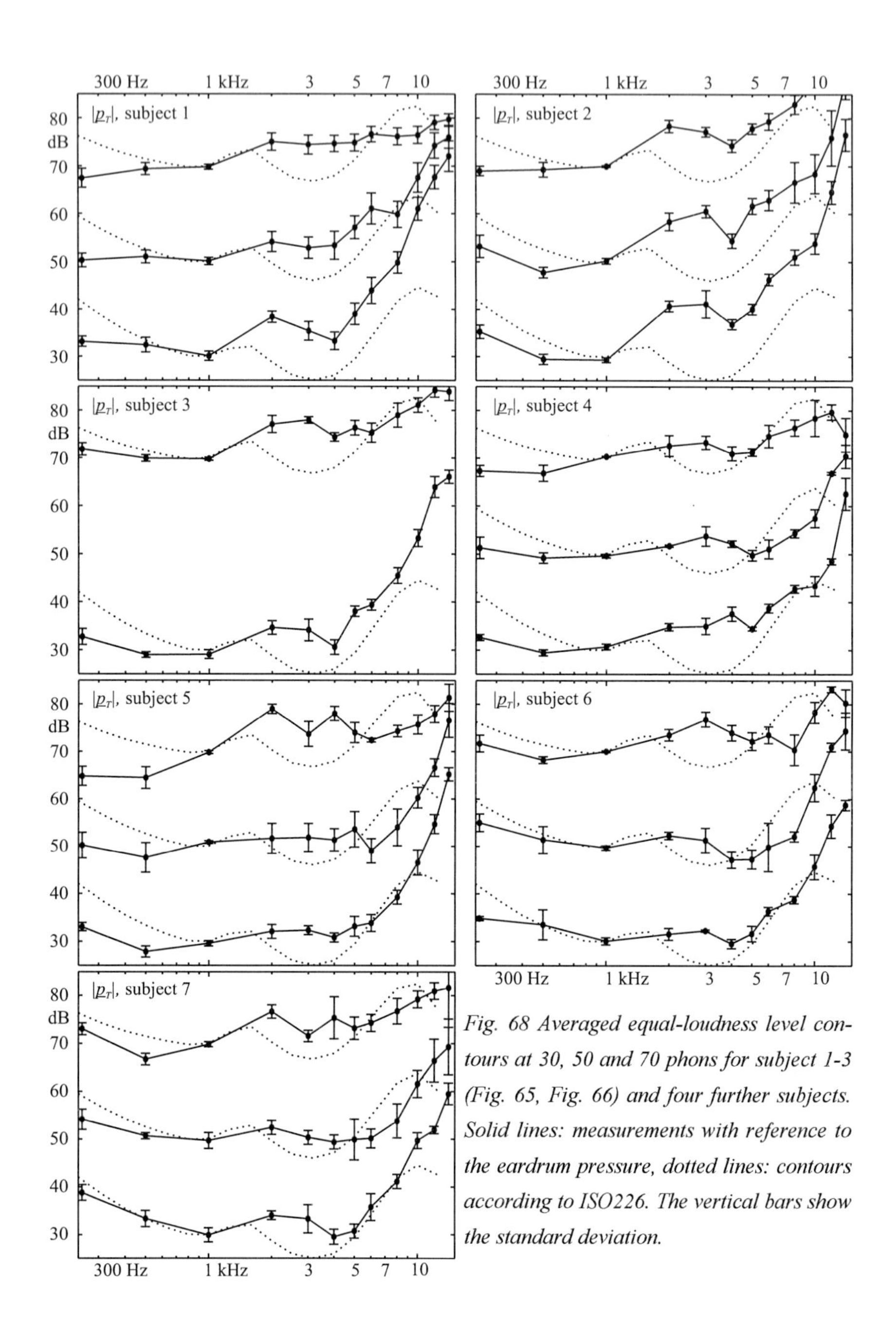

Fig. 68 Averaged equal-loudness level contours at 30, 50 and 70 phons for subject 1-3 (Fig. 65, Fig. 66) and four further subjects. Solid lines: measurements with reference to the eardrum pressure, dotted lines: contours according to ISO226. The vertical bars show the standard deviation.

of eardrum SPL and approaches the ISO 226 curve (the individual contour of subject 1 was not expected to match the standard exactly). These results confirm the findings of Gabriel et al., 1997: the low-frequency deviations between the current measurements and the ISO standard are caused by the choice of the initial levels. As this is not relevant for higher frequencies, the low frequency reference level of 30 dB SPL was retained in the following experiments in spite of the deviations.

3.3.4.3 Interindividual comparison

After it was shown that the measured contours do not depend on the probe microphone position and that the low-frequency deviations are generated by the selection of the initial sound pressure level, a feature comparison of individual curves is carried out.

To extend the set of subjects, the equal-loudness level contours at 30, 50 and 70 phons were measured with four further listeners (subjects 4-7). The microphone position during the estimation measurements depended on the shape of the particular probe pressure. The resulting frequencies of the first minimum were located between 4400 and 8700 Hz. Exclusively second-order estimations were applied.

The equal-loudness level contours measured with relation to the eardrum pressure do not show minima that can be distinctly attributed to ear canal resonances (Fig. 68). Generally, all data points near the first canal minimum are located approximately 10 dB above the standard curves as a result of the ear canal transfer function compensation. At higher frequencies, the measured curves increase (except for the 70 phons curves of subjects 3, 4 and 5 at 16 kHz) which suggests that the second minimum is compensated as well. However, between 1 and 4 kHz, most curves are not regularly shaped. Instead of a monotonous slope, a peak at 2 kHz that is followed by a notch at 4 kHz can be observed. This is particularly distinctive in the contours determined for 30 phons. As the applied frequency resolution in not sufficient to investigate the shape of the curves exactly, an additional measurement was carried out.

3.3.4.4 Increased frequency resolution of the contour at 30 phons

In this experiment, exclusively the contour at 30 phons is determined because it shows the most distinct shape features between 1 and 5 kHz. As similar shapes can be found in the contours of each subject, the measurement was only carried out with one listener (subject 1).

The frequencies were increased in third octave steps (according to ISO 226), starting at 500 Hz. According to the previous experiments, exclusively second-order estimation was applied. Four measurements were averaged for each frequency. The results are depicted in Fig.

69 in addition with the data points already shown in Fig. 67 to continue the contour for low frequencies (30 dB initial SPL, reference to $|\underline{p}_T|$). Furthermore, the results of the preliminary experiment which was carried out 8 months earlier are displayed (gray circles).

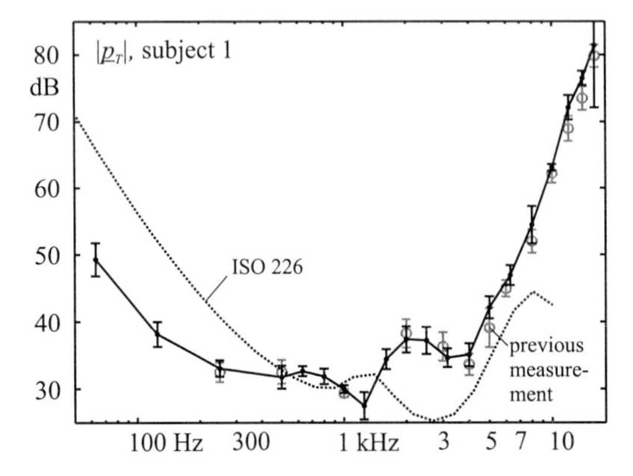

Fig. 69: Equal-loudness level contours for subject 1, 30 phons. Above 500 Hz, third octave frequency resolution was utilized. For comparison, the average of the previous measurement (Fig. 65) is depicted in gray. Vertical bars denote the standard deviation.

Obviously, the recent measurement is in good agreement with the previously determined contours. The maximal deviation between the experiments is approximately 3 dB at 5 kHz. The other data points are located within the standard deviation, except for very high frequencies. The shape of the contour between 1 and 4 kHz shows the expected irregular shape. At 1250 Hz, a notch is visible, while the data points between 1600 Hz and 3150 Hz are significantly raised.

3.3.5 Discussion

The preliminary experiment shows that the difference between equal-loudness level contours basing on estimation signals from different positions in the ear canal has the same order of magnitude as the standard deviation of the measurements. In conclusion, contours determined with relation to the eardrum pressure do not depend on the insertion depth of the estimation probe microphones.

The results of the low-frequency analysis support the empirical findings of Gabriel, 1997. The deviations for low frequencies are based on range effects implied by the initial sound pressure level. Low-frequency results do not depend on whether the reference signal is specified at the tympanic membrane or in the free sound field. As range effects are not a main issue of this investigation, the primarily implemented experimental design (no adaptation of the initial SPL for low frequencies) was not changed in the remaining measurements.

The individual measurements of seven subjects show some general features of contours determined with reference to the eardrum sound pressure. It becomes obvious that the curves are free of the ear canal minimum between 2 and 4 kHz and of the decline that can be found in the standard contours for higher frequencies. Thus, averaging the contours would be more meaningful than averaging over curves containing the canal minima. However, a notch (1.25 kHz) and a broad peak (2 kHz) of 5-10 dB can be observed. For all subjects, this shape is clearly visible in the contour at 30 phons and becomes less distinctive for higher values of loudness. As the measurement is related to the eardrum sound pressure, the ripples cannot be generated by duct effects of the ear canal. Hence, they must be assigned to the middle or inner ear or to higher levels of sound perception.

The irregular shape of the curve suggested a new investigation of the 30 phons contour with a higher frequency resolution. Measurements with frequency points in third octave steps confirmed the observed structure of the curve between 1 and 4 kHz. Similar contour shapes are published in other studies (Killion, 1978; Puria et al., 1997). The contour shown in Fig. 69 is remarkably similar to the absolute hearing threshold estimates documented by Killion, except for the notch at 1.25 kHz. In the reference, various measurements of monaural hearing thresholds were adjusted for the individual ear canal influence and combined. The reported dip in sensitivity (which is equivalent to the broad peak of the equal-loudness level contour at 2 kHz) is almost exactly reproduced by the present investigation, although the curve at 30 phons was determined instead of the hearing threshold. Unfortunately, the origin of the "hump" remains unclear. However, the proposed measurement method would significantly enhance future investigations on this issue.

4 Summary and outlook

4.1 Summary

In this thesis, an efficient and accurate method for the estimation of the sound pressure at the human eardrum was proposed and evaluated. The eardrum signal is well suited as a reference for measurements at the ear. Most traditional approaches for the determination of the eardrum pressure are based on the adaptation of a physical model of the ear canal to previously measured individual parameters. The ear canal can be roughly modeled as an inhomogeneous acoustical duct that is terminated by the impedance of the eardrum. Using circuit models of the canal that can be derived from the canal geometry, sound field values (pressure, volume velocity, impedance) can be transformed from the entrance to the tympanic membrane. Often, inverse techniques are used to derive the ear canal geometry from input impedance measurements. With a previous implementation of this method (Hudde et al., 1999), accurate results were obtained in artificial inhomogeneous, but straight ear canals. However, the algorithm often failed in natural ear canals. The absence of curvature in artificial dummy canals suggested that spatial sound field effects occurring in the curves and bends of real canal geometries are responsible for the estimation errors. Additionally, spatial effects at the orifice of the impedance measurement device which is connected to the ear canal can be expected. Practically, a local area discontinuity cannot be avoided. Thus, sound waves traveling through the orifice excite evanescent higher modes which were expected to distort the impedance measurement results. To improve the method, two main questions had to be answered:

- Does the spatial sound field allow for a shape determination of the ear canal with data measured at the canal entrance? If not, is it possible to estimate the signals at the eardrum directly from the entrance?

- Are impedance measurements at the ear canal entrance sufficiently accurate to solve the estimation problem?

Obviously, it was necessary to examine the spatial sound field structure for various frequencies. For a three-dimensional representation, a Finite Element (FE) model of the sound fields arising at the pinna and in the ear canal for different sources was implemented. The pinna was digitized from magnetic resonance images of a human subject. Several ear canal models were designed that agree with natural canal shapes regarding essential geometry features. Tympanic membrane and middle ear were extracted from an existing FE model of the human head. The simulated sound field structures were visualized by means of equidistant isosurfaces (surfaces of constant pressure magnitude).

The results of the simulation show that the sound field can be subdivided into three parts: (a) the eardrum coupling region near the tympanic membrane, (b) the fundamental sound field in the core region of the canal and (c) the distinctly three-dimensional external part which lies outside of the entrance surface. This configuration complies with the usual one-dimensional modeling approaches. However, the three-dimensional simulation provided insight into some important details.

In the eardrum coupling region, the sound field is essentially influenced by the eardrum. The models show that the sound waves are guided towards the point **T** into the tympanomeatal angle. The tympanic membrane thus acts as continuation of the canal walls. When the elastic tympanic membrane is replaced by a rigid eardrum, differences are only visible at the main eardrum and middle ear resonance for frequencies between 600 Hz and 4 kHz. These differences are generated by the near field of eardrum sound radiation. However, due to the relatively large wavelength, the pressure variations in the vicinity of the tympanic membrane are very small for the mentioned frequencies. This finding supports lumped element models of the eardrum coupling region. At the point **T**, the maximal ear canal sound pressure occurs practically in the whole hearing frequency range. Thus, the spectrum of p_T is flat in comparison to other ear canal signals. It is reasonable to specify p_T as the desired eardrum signal ("input to the middle ear").

In anterior direction, the ear canal is continued by the core region in which the fairly regular fundamental sound field of the canal can be found. The fundamental field is independent of the sound source. It mainly consists of slightly curved isosurfaces that are aligned perpendicular to the canal walls and to a virtual middle axis connecting the centroids of the isosurfaces. The sound field characteristics vary only along one curvilinear coordinate, thus, it is often characterized as unidimensional. The external sound field in front of the pinna and in the *cavum conchae* obtains distinctly three-dimentional isosurface structures, which merge with the fundamental field in the anterior section of the canal. Hence, the entrance of the core region can be specified as the most anterior sound field isosurface which is independent from the source. However, orientation and position of the core region entrance surface depend on frequency. In particular, local pressure extrema can disturb its shape. Pressure minima result in large velocities which generate a strong coupling of the canal sound to the external field. Consequently, the sound source has no significant influence on the isosurface shape in the anterior canal section. When a pressure maximum arises at the ear canal entrance, the fields are coupled only weakly, as a velocity minimum is present. Thus, the position of the first source-independent isosurface is shifted towards the tympanic membrane.

Particularly in strongly curved sections of the core region, "one-sided" isosurfaces occur which are domed over a point on the canal wall. In general, one-sided isosurfaces arise over pressure extrema that occur on concave boundaries. It was shown that the sound field in the ear canal maintains minimum energy density in spite of the present one-sided isosurfaces. The disturbance of the regular sound field is restricted to a region of minimal spatial extension. Furthermore, the energy density is raised above the minimal level only, when the pressure magnitude gradient and the velocity vectors have different orientation. This situation only occurs for a short time interval of the oscillation cycle. As a consequence, one-sided isosurfaces of sound pressure magnitude are not identical with the local phase isosurfaces. Fundamental mode propagation is characterized by minimal energy density as well. However, it is impossible to decompose the sound field in the core region into distinct propagation modes that do not depend on frequency. Thus, it is referred to as "fundamental sound field".

In the sections of the fundamental sound field in which regular isosurfaces prevail, magnitude and phase isosurfaces are coincident and do only weakly depend on frequency. However, as one-sided isosurfaces are arranged concentrically around pressure magnitude extrema, their position varies with frequency. The definition of a middle axis by one-sided magnitude isosurfaces is critical. Fortunately, the phase isosurfaces are regular even in the vicinity of magnitude extrema. In the core region, the phase isosurfaces are independent of frequency and are identical for incident and reflected waves. Hence, phase isosurfaces can be interpreted as indicators of wave front propagation. When the middle axis is defined according to phase isosurfaces, is is also independent of frequency. It remains meaningful in regions of one-sided magnitude isosurfaces as well, because the local velocity vectors are aligned perpendicular to the phase isosurfaces for the most of the time.

One-sided isosurfaces are effects of local evanescent higher order modes. It turned out that the transfer function over irregular regions is not significantly different from the straight duct case. However, measurement equipment should not be attached to the canal in regions of irregular isosurfaces, as the local field is highly sensitive to disturbances. The simulations carried out for the external ear show that the entrance surface of the core region is typically located behind the first bend of the ear canal, as the field structure does not depend on the excitation direction of the model here.

After the basic characteristics of the external ear sound field were investigated, the application of measurement devices that are attached to the ear canal by coupling tubes was examined. The measurement of acoustical impedances at the canal entrance requires a robust and well-specified, yet practically feasible coupling to the canal. It is necessary to attach the measurement device in the core region of the canal. In doing so, the entrance surface of the core region must not be altered. Below approximately 4 kHz, the ear canal mainly behaves as lumped element due to the relatively large wavelength. For higher frequencies, acoustical impedance is essentially determined by the cross-sectional area function of the canal, in particular the frequencies of minima and maxima. When the ear canal shape is calculated by the impedance function, it is crucial to determine the impedance extrema correctly. At the orifice of the coupling tube, an unavoidable area discontinuity occurs. It was expected that evanescent higher order modes have noticeable impact on the local sound field. Thus, FE models of typical measurements were implemented. From a reference model, two versions with altered connecting tube radius and two further models with inclined tubes were derived to study the effects separately. By visualization of the arising sound fields, the influence of the connecting tubes on the entrance of the core region could be examined in detail as well.

The model calculations show that the extremal frequencies are significantly affected by the coupling of the tube and the canal, in particular the minima. The deviation of the minima depends on the area discontinuity. It was not possible to obtain a satisfying compensation of the errors using equivalent elements that model the influence of higher order modes (Karal, 1953). The results suggested that optimal coupling is obtained by a measurement device that matches the cross-sectional area of the entrance surface of the canal. The impedance maxima are not significantly altered by the area discontinuity, but depend on the orientation of the tube with respect to the middle axis of the ear canal. These errors are smaller than the deviations of the minima that are generated by the area discontinuity.

The analysis of the sound field structure arising in the simulated ear canals and connecting tubes obtains similar results as the examination of the sound field coupling of the concha and the ear canal: when pressure minima (which correspond to velocity maxima) occur at the orifice, tube and ear canal are strongly coupled. As the velocity vectors radially extend into the ear canal, the effective length of the canal is increased. Thus, the frequencies of the corresponding pressure minima are decreased. The sound field coupling is low for pressure minima: as the velocity coupling is very small, the exact position of the pressure maximum is shifted easily by altering the tube orientation. Generally, the presence of the tube orifice alters the shape of the adjacent isosurfaces significantly. Thus, the local field depends on the sound source. The entrance area of the core region is shifted to more posterior positions by the connecting tube. Strictly spoken, it is impossible to carry out impedance measurements within the core region unless a connecting tube is used that leaves the isosurface shape occurring in the free ear canal unaltered. Such devices are not practically feasible.

In conclusion, all measurement methods and calculations that are based on the concept of constant scalar sound field variables on fixed areas at the ear canal entrance underly the problems that were observed for the impedance measurement models.

Impedance measurements at the ear canal entrance are meaningful for frequencies below 3-4 kHz. Measurements at higher frequencies are basically influenced by the characteristics of the coupling of the device and the ear canal. It is not recommended to use such data for ear canal geometry estimation, because the necessary extremal frequencies have to be known with high accuracy. When geometries determined by inaccurate impedance data are transformed into one-dimensional models, additional errors can be expected due to the distinctly three-dimensional structure of the ear canal sound field.

These findings provide some orientation for the design of the eardrum pressure estimation algorithm. Duct models based on the ear canal geometry should be avoided, because impedance measurements are not suited to deliver accurate data for an estimation of the ear canal shape. Thus, the transfer function between the measurement microphone and the tympanic membrane is estimated directly. As point transfer functions are independent of the source even at small distances in front of the entrance surface of the core region, probe microphones that implement point measurements in good approximation are used for the necessary measurements.

To estimate the transfer function of the ear canal sound field, the chain parameters of an equivalent duct model are adapted to the minima in the measured pressure spectrum. Above 3-4 kHz, the eardrum has sufficiently high pressure reflectance to generate standing wave minima in the canal. For typical probe insertion depths, the first two minima of the pressure occur in the frequency range below 20 kHz. The entrance impedance minima of the equivalent duct model are adjusted to match the respective probe pressure minima by numerical adaptation of the duct length and shape (cylindrical or conical). Propagation losses in the ear canal and energy absorption by the tympanic membrane are modeled by a lumped resistance element terminating the equivalent duct. In the developed software implementation, the element is adjusted manually to obtain a maximally flat eardrum sound pressure. Example calculations show that the estimation error remains surprisingly small even for low power reflectance values of the ear canal termination. For a robust detection of the required minima, successive instantaneous spectra of the microphone pressure are subtracted, while the microphone is inserted into the canal. The maximal changes that are identified this way are identical with the required minima.

The method is highly feasible, as only one pressure probe is necessary and works fast enough to be applied immediately before psychoacoustical experiments. The estimation technique was evaluated using finite-element models of the external ear and an artificial ear. Both methods yield comparable results. It was possible to determine the eardrum sound pressure magnitude within error limits of 2-3 dB. The maximum error arises at the minima used for estimation, for intermediate frequencies, the error is substantially smaller. As long as the measurements are carried out in the fundamental sound field, the accuracy does not depend on the probe tip position in the ear canal. Furthermore, influences of the ear canal shape, length and eardrum orientation could not be observed.

It is the preferential application of the estimation method to provide a well-specified reference for psychoacoustical measurements. To evaluate the practical feasibility, equal-loudness level contours with relation to the eardrum pressure were determined. In addition, the results serve as indirect evaluation of the method. For this reason, a preliminary experiment was carried out to determine, if the probe position influences the estimation results. A dependence of the contours on the position of the estimation microphone is hardly visible. Measurements of equal-loudness level contours on several subjects show that the individual ear canal influence is minimized, if the estimated eardrum pressure is taken as reference signal.

4.2 Outlook

Using the described technique, an efficient and practically feasible method for eardrum pressure estimation is available. It is expected that many varieties of psychoacoustical experiments and audiological applications can benefit from using p_T as reference signal. This suggests a broad range of possible succeeding investigations. In this outlook, only some examples can be given:

- *Standardization of equal-loudness level contours*: The cancellation of individual ear canal transfer functions is one of the main issues for eardrum pressure reference. In the work at hand, it was not intended to determine equal-loudness level contours for a large number of subjects. Nevertheless, a new standard of equal-loudness level contours with relation to the tympanic membrane would be of great benefit for auditory research and applications. Thus, a thorough study of contours gathered from a significantly large number of listeners would be desirable.
- *Psychoacoustical measurements with increased accuracy*: The ear canal transfer function does not only alter the magnitude of source pressure signals, it has essential influence on the phase angle as well. The phase response of the ear canal is unimportant in most measurements using pure tones (such as the experiments documented in this thesis). However, in numerous other experiments, broad-band signals are applied. By means of the estimation method, stimuli can be delivered to the ear without excessive phase distortion by the ear canal. It is thus reasonable, to apply the method to experiments in which the phase angle of the stimulus is important (e.g. measurements using Schroeder phase signals).
- *Individualized eardrum related audiometry*: In standardized hearing threshold measurements, a frequency dependent headphone signal according to the ISO 226 equal-loudness level contours is used. Audiometers are calibrated by sound pressure measurements in headphone couplers. Thus, interindividual differences of the external ear transfer function significantly influence the determined hearing threshold. If the eardrum pressure is selected as reference, comparable audiometry results can be achieved.
- *Measurement of hearing aid gain with respect to the eardrum*: Probe microphone measurements can be carried out while the earmould or the receiver of a hearing aid is placed inside the ear canal. The results can be adapted to eardrum related audiometry data to improve hearing aid fitting. Possibly, microphones that are attached to the earmould and directed into the ear canal can be used to estimate the hearing aid gain. However, problems due to the higher order modes generated by the area discontinuity at the hearing aid orifice may arise. This issue can be investigated in successive studies using models similar to the measurement simulations discussed in section 2.5.

A significant extension of the method is possible, if a generalized middle ear transfer function is added. Such transfer functions could be derived from a middle ear network model (e.g. Hudde and Engel, 1998abc). The eardrum sound pressure could be transformed to the *stapes* volume velocity or the pressure in the vestibule. However, middle ears underly interindividual variations as well. Probably, the generalized middle ear model could be adapted to features of p_T to approximate the individual middle ear transfer function.

In other applications, the transfer function between the tympanic membrane and points at the ear canal entrance is necessary, for instance when otoacoustic emissions (OAEs) are to be evaluated. It would be favorable to estimate the transfer function by the proposed method, in particular, as both sound source and probe microphone can be left at the ear canal entrance and reused for the OAE measurement. The effective reverse transfer function is unfortunately not equivalent to the inverse of the estimated function, because the ear canal is not symmetric and terminated differently at its ends. Further, the shape of eardrum vibrations is different in both cases. Nevertheless, it is most likely that a method for the accurate estimation of the reverse transfer function can be developed using the given external ear model.

Appendix

A.1 The linear acoustical wave equation, duct acoustics and chain matrices

Sound waves propagating in a fluid medium can be considered as variation of density which depends both on position and on time. The temporal and spatial variations are coupled by partial differential equations, which can be set up by three basic terms, the *state equation*

$$p = c^2 \rho \tag{A.1.1}$$

the *Euler equation* (given in one and three dimensions)

$$\frac{\partial p}{\partial x} = -\rho \frac{\partial v_x}{\partial t}; \qquad \mathrm{grad}(p) = -\rho \frac{\partial \mathbf{v}}{\partial t} \tag{A.1.2}$$

and the *continuity equation*

$$\frac{\partial v_x}{\partial x} = -\frac{1}{\rho c^2} \frac{\partial p}{\partial t}; \qquad \mathrm{div}(\mathbf{v}) = -\frac{1}{\rho c^2} \frac{\partial p}{\partial t} \tag{A.1.3}$$

The quantities p, v and ρ denote the sound pressure, velocity and density variation in the fluid. The values ρ and c represent the static density and the speed of sound waves. The three equations are functions of the variables x (spatial coordinate) and t (time)

When the equations (A.1.1) to (A.1.3) are coupled by partial differentiation with respect to the spatial variables and time, the three-dimensional wave equation can be set up for the sound pressure:

$$\ddot{p} = c^2 \Delta p \tag{A.1.4}$$

Here, the character Δ represents the Laplace operator in Cartesian coordinates:

$$\Delta f = \frac{\partial^2 f}{\partial x^2} + \frac{\partial^2 f}{\partial y^2} + \frac{\partial^2 f}{\partial z^2} \tag{A.1.5}$$

The field geometry depends on the boundary conditions of the regarded region. For generally shaped volumes, usually no analytic solution of the wave equation can be found.

In one-dimensional coordinates, the solution of the wave equation can be obtained by the superposition of incident and reflected waves, as can be easily verified by substitution:

$$p_x(x,t) = p_i(x - ct) + p_r(x + ct) \tag{A.1.6}$$

This equation is called *d'Alembert solution*. When it is applied to harmonic functions (sinusoidal signals) that are represented by complex phasors, the *Bernoulli solution* results:

$$\frac{\partial^2 \underline{p}}{\partial x^2} + \beta^2 \underline{p} = 0 \tag{A.1.7}$$

The wave number β is defined as $\beta = \omega/c = 2\pi/\lambda$. Two important analytic solutions of the wave equation for special boundary conditions are plane wave propagation

$$\underline{p}(x,\beta) = p_0 \cdot e^{-j\beta x}; \quad \underline{v}(x,\beta) = \frac{p_0}{Z_F} \cdot e^{-j\beta x}; \quad Z_F = \rho c \tag{A.1.8}$$

and zeroth order spherical wave propagation

$$\underline{p}(r,\beta) = \frac{p_0}{r} \cdot e^{-j\beta r}; \quad \underline{v}(x,\beta) = \frac{p_0}{r\underline{Z}_F} \cdot e^{-j\beta r}; \quad \underline{Z}_F = \frac{\rho c}{1 + \frac{1}{j\beta r}} \tag{A.1.9}$$

The equations describe exclusively the forward traveling wave according to equation (A.1.6). The variable $\underline{Z}_F$ represents the field impedance, which relates pressure and velocity in a sound field. The pressure p_0 denotes the maximum pressure in the wave. The wave propagation direction is oriented along the length or radius coordinates x and r, respectively.

Planar wave propagation occurs in cylindrical ducts that are excited by a source vibrating parallel to the middle axis of the tube. Usually, the sound velocity $\underline{v}$ in the duct is related to its cross-sectional area A and replaced by the volume velocity $\underline{q}$ as field variable:

$$\underline{q} = \int_A \underline{\mathbf{v}}\, d\mathbf{A} = j\omega \underline{V} \tag{A.1.10}$$

The sound pressure divided by the volume velocity on a specified surface is defined as acoustical impedance $\underline{Z} = \underline{p}/\underline{q}$. On surfaces with infinite acoustical impedance, the volume velocity vanishes and the pressure becomes maximal (rigid boundary). When a duct is terminated perfectly rigid, incoming pressure waves are reflected without change of sign. A termination impedance that is equivalent to the tube wave impedance yields a matched boundary, where no power reflection occurs. When the termination impedance is zero, incoming pressure waves are reflected completely with changed sign.

Using the *d'Alembert solution* (A.1.6), the plane wave solution (A.1.8) and the volume flow integral (A.1.10), the sound field variables at different positions in a duct ($x_1 = 0;\ x_2 = \Delta L$) can be related. It holds

$$\underline{p}_1 = \underline{p}(x_1) = p_i + p_r,\quad \underline{p}_2 = \underline{p}(x_2) = p_i e^{-j\beta\Delta L} + p_r e^{j\beta\Delta L} \tag{A.1.11}$$

$$\underline{q}_1 = \underline{q}(x_1) = \frac{p_i}{Z_{tw}} - \frac{p_r}{Z_{tw}},\quad \underline{q}_2 = \underline{q}(x_2) = \frac{p_i}{Z_{tw}} e^{-j\beta\Delta L} - \frac{p_r}{Z_{tw}} e^{j\beta\Delta L} \tag{A.1.12}$$

Equation (A.1.12) can be deduced by applying the *Euler equation* to equation (A.1.11). Due to the transition from the velocity $\underline{v}$ to the volume velocity $\underline{q}$, the tube wave impedance $Z_{tw}=\rho c/A$ has to be used instead of the field impedance Z_F. Using basic theorems of trigonometry, the above equations can be rewritten as chain matrix of the duct between the two positions x_1 and x_2:

$$\begin{pmatrix} \underline{p}_1 \\ \underline{q}_1 \end{pmatrix} = \begin{pmatrix} \underline{C}_{11} & \underline{C}_{12} \\ \underline{C}_{21} & \underline{C}_{22} \end{pmatrix} \begin{pmatrix} \underline{p}_2 \\ \underline{q}_2 \end{pmatrix} = \begin{pmatrix} \cos(\beta\cdot\Delta L) & jZ_L \sin(\beta\cdot\Delta L) \\ j/Z_L \sin(\beta\cdot\Delta L) & \cos(\beta\cdot\Delta L) \end{pmatrix} \begin{pmatrix} \underline{p}_2 \\ \underline{q}_2 \end{pmatrix} \tag{A.1.13}$$

The chain matrix of a cylindrical duct models its transfer characteristics exactly, as far as only plane wave propagation occurs. In other words, only a fundamental mode sound field is present. The cross-sectional areas bounding the respective region can be regarded as ports of a one-dimensional transmission line model in which only scalar signals $\underline{p}$ and $\underline{q}$ occur. The junction of cylindrical ducts with different cross-sectional areas (and hence different tube wave impedances) can be represented by consecutive multiplication of the respective chain matrices, as pressure and volume flow at the output of one segment are equal to the input values of the next. By this method, complicated duct structures can be approximated using stepped ducts. However, as higher order modes occur near the junction between two different tube segments, the model is subject to essential errors.

In conical ducts, the fundamental mode is represented by spherical wave propagation. Again, the *d'Alembert solution* (A.1.6) can be used to set up the duct equations in conjunction with the spherical wave solution (A.1.9). As the conical tube is less symmetric than the cylindrical case (the tube wave impedance depends on the position on the middle axis), the resulting chain matrix becomes more complicated.

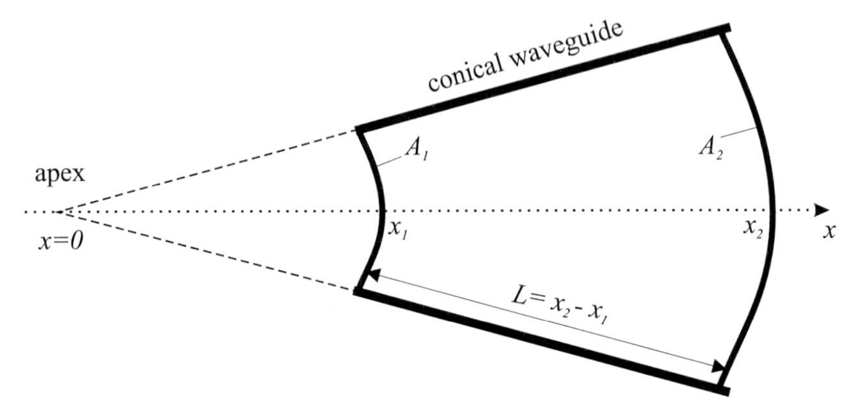

Fig. A.1.1: Geometry reference for the conical waveguide

Fig. A.1.1 shows the specification of the duct geometry and the necessary variables. The chain parameters are given as

$$\begin{pmatrix} \underline{p}_1 \\ \underline{q}_1 \end{pmatrix} = \begin{pmatrix} \underline{F}_1 \cdot e^{j\beta L} + \underline{F}_2 \cdot e^{-j\beta L} & \underline{F}_1 \cdot \underline{Z}_{2r} \cdot e^{j\beta L} - \underline{F}_2 \cdot \underline{Z}_{2i} \cdot e^{-j\beta L} \\ \underline{F}_1 / \underline{Z}_{1i} \cdot e^{j\beta L} - \underline{F}_2 / \underline{Z}_{1r} \cdot e^{-j\beta L} & \underline{F}_1 \cdot \underline{Z}_{2r} / \underline{Z}_{1i} \cdot e^{j\beta L} + \underline{F}_2 \cdot \underline{Z}_{2i} / \underline{Z}_{1r} \cdot e^{-j\beta L} \end{pmatrix} \begin{pmatrix} \underline{p}_2 \\ \underline{q}_2 \end{pmatrix} \tag{A.1.14}$$

with

$$\underline{F}_1 = \frac{x_2}{x_1} \cdot \frac{1}{1 + \underline{Z}_{2r} / \underline{Z}_{2i}}; \quad \underline{F}_2 = \frac{x_2}{x_1} \cdot \frac{1}{1 + \underline{Z}_{2i} / \underline{Z}_{2r}} \tag{A.1.15}$$

and

$$\underline{Z}_{1,2i} = \frac{\rho c}{A_{1,2}} \cdot \frac{1}{1 + \dfrac{1}{j\beta x_{1,2}}}; \quad \underline{Z}_{1,2r} = \frac{\rho c}{A_{1,2}} \cdot \frac{1}{1 - \dfrac{1}{j\beta x_{1,2}}} \tag{A.1.16}$$

The four impedances defined by equation (A.1.16) can be interpreted as tube wave impedances occurring at the two port areas for the incident and reflected wave, respectively.

Chain matrices can be used to transform acoustical impedances between two positions in a duct or to compute transfer functions, when the termination impedance is known. With the set of equations (A.1.13), the impedance $\underline{Z}_X$ at a distance of ΔL from a termination with $\underline{Z}_T$ can be expressed as

$$\underline{Z}_X = \frac{\underline{p}_1}{\underline{q}_1} = \frac{\underline{C}_{11}\underline{Z}_T + \underline{C}_{12}}{\underline{C}_{21}\underline{Z}_T + \underline{C}_{22}} \tag{A.1.17}$$

By inversion of (A.1.17), the termination impedance of a duct with known chain parameters can be calculated from impedance measurements at its entrance. The pressure transfer function over the duct can be calculated as

$$\underline{H} = \underline{C}_{11} + \underline{C}_{12} / \underline{Z}_T \tag{A.1.18}$$

Alternatively, the termination impedance can be calculated from the transfer function measured over a duct with known chain parameters.

A.2 Material parameters

The material parameters of air (speed of sound and mass density) were set to c=350 m/s and ρ=1.18 kg/m³ in all FE calculations, except where indicated otherwise.

The mass density of the tympanic membrane was set to ρ=1200 kg/m³. It consists of two different materials modeling the *pars flaccida* and *pars tensa*, thus the Young's moduli were adjusted to E_X=1.0 MPa *and* E_X=2.1 MPa. Poisson's ratio was set to 0.4. The β damping constant defines the factor by which the stiffness matrix is multiplied to obtain the damping matrix of the FE system. For the tympanic membrane, it was represented by a value of 10 μs.

The material of the ossicles yields a Young's modulus of 10^{10} Pa, hence they are modeled to be approximately rigid. The applied density of ρ=2000 kg/m³ results in masses of 21.5 mg for the *malleus*, 30.4 mg for the *incus* and 3.46 mg for the *stapes*.

	Translational compliance	Rotational compliance
malleal ligament	0.2 mm/N	$5\ N^{-1}mm^{-1}$
incudal ligament	0.2 mm/N	$5\ N^{-1}mm^{-1}$
incudomalleal joint	$5\cdot10^{-3}$ mm/N	$0.75\cdot10^{-3}\ N^{-1}mm^{-1}$
incudostapedial joint	$7\cdot10^{-2}$ mm/N	$5\ N^{-1}mm^{-1}$
annular ligament	0.8 mm/N	-

The joints and ligaments are modeled as cylindrical equivalent elements. Weistenhöfer, 2002, developed a model describing all the elastic elements by translational and rotational compliances and resistances. In the direction of the main axis of operation, the equivalents yield the following parameters:

The hydroacoustical load on the *stapes* is approximated as a mechanical second-order vibrator (mass 10 mg, compliance 11 N/mm, mechanical resistance 70 Ns/mm) according to Hudde and Engel, 1998abc.

A.3 Measurement hardware setup

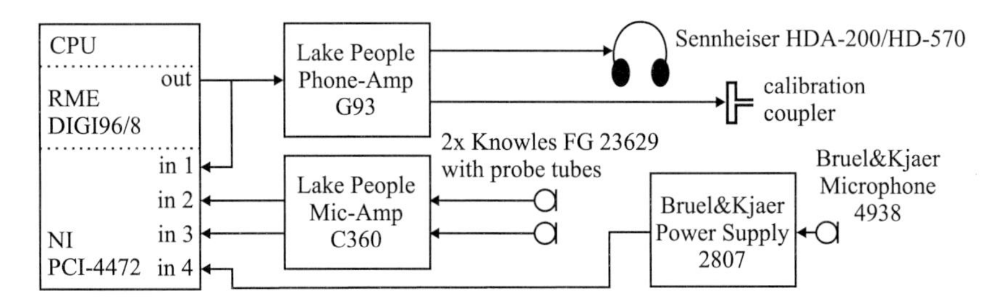

Fig. A.3.1: Schematic drawing of the measurement hardware

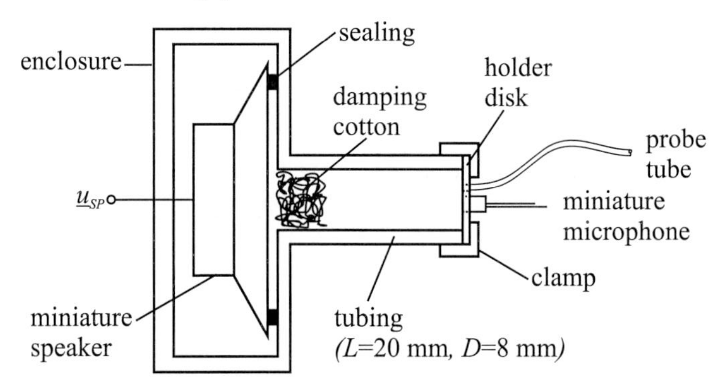

Fig. A.3.2: Schematic drawing of the calibration coupler

A schematic drawing of the data acquisition equipment is given in Fig. A.3.1. Signals were digitized by a National Instruments® PCI-4472 measurement card. For probe microphone voltage supply and preamplification, devices with suitable signal-to-noise and distortion characteristics were selected (Lake People® Mic-Amp C360: 129 dB SNR at 60 dB gain, less than 0.001% THD). As calibration reference microphone, a B&K® 4938 condenser capsule with a B&K® 2807 power supply was applied. Sennheiser® HDA-200 headphones were used as source. The headphone amplifier (Lake People® Phone-Amp G93) is automatically adapted to the electrical impedance of the headphones or the loudspeaker in the utilized calibration coupler. The coupler for microphone calibration consists of a miniature loudspeaker with attached brass tubing (see Fig. A.3.2). Various metal disks can be mounted to the end of the tube and take the B&K® reference microphone, the tip of the probe tube microphones or the Knowles® FG 23629 capsule without attached probe. The disks are fabricated to hold each of the microphones in the same position of the coupler tube. Thus, the coupler pressure can be assumed as constant.

A.4 Moving average implementation

To increase the signal-to-noise ratio of measurements, several instantaneous measurements can be averaged. In this context, "instantaneous measurement" means a time-dependent signal x observed at the time instant n. The signal x may be a time domain signal as well as an instantaneous spectrum as a function of frequency. When spectra are averaged, their empirical expectancy value is calculated.

The moving average y of several observances at the time instant n is given as

$$y_n = \frac{x_n + x_{n-1} + \ldots + x_1}{n} \tag{A.4.1}$$

Unfortunately, all previous measurements x_1 to x_{n-1} have to be stored in addition to the present measurement x_n. In practice, the calculation should be carried out recursively:

$$ny_n = x_n + \ldots + x_1 = x_n + (n-1) \cdot \left(\frac{x_{n-1} + \ldots + x_1}{n-1} \right) = x_n + (n-1) \cdot y_{n-1} \tag{A.4.2}$$

and

$$y_n = \frac{x_n + (n-1) \cdot y_{n-1}}{n} \tag{A.4.4}$$

Samples from recent time instants are often more significant than earlier measurements. To increase the influence of recent data, moving window functions can be implemented. For instance, exponential functions are used as window:

$$w_n = \alpha^n \quad \forall\ \mathrm{n} \geq 0, 0 \leq \alpha < 1 \tag{A.4.5}$$

Equation (A.4.1) becomes

$$y_n = \frac{x_n w_0 + x_{n-1} w_1 + \ldots + x_1 w_{n-1} + x_0 w_n}{w_0 + w_1 + \ldots + w_{n-1} + w_n} = \frac{\sum_{k=0}^{n} x_k w_{n-k}}{\sum_{k=0}^{n} w_k} = \frac{\sum_{k=0}^{n} x_k \alpha^{n-k}}{\sum_{k=0}^{n} \alpha^k} \tag{A.4.6}$$

The variable α represents the memory time constant of the averaging system. The smaller α is chosen, the smaller number of previous measurements are taken into account. With $\alpha=1$, equation (A.4.6) becomes (A.4.1). It can be transformed into a recursive formulation as well:

$$y_n \sum_{k=0}^{n} \alpha^k = \sum_{k=0}^{n} x_k \alpha^{n-k} = x_n + \sum_{k=0}^{n-1} x_k \alpha^{n-k} \tag{A.4.7}$$

The sum can be expressed as function of y_{n-1}:

$$y_n \sum_{k=0}^{n} \alpha^k = x_n + \left(\sum_{k=0}^{n-1} \alpha^k\right) \frac{\alpha \sum_{k=0}^{n-1} x_k \alpha^{(n-1)-k}}{\sum_{k=0}^{n-1} \alpha^k} = x_n + \alpha \left(\sum_{k=0}^{n-1} \alpha^k\right) y_{n-1} = x_n + \left(\sum_{k=1}^{n} \alpha^k\right) y_{n-1} \tag{A.4.8}$$

Using the geometric sum

$$\sum_{k=0}^{n} \alpha^k = \frac{1-\alpha^n}{1-\alpha} \quad \forall\, n \geq 0, 0 \leq \alpha < 1 \tag{A.4.9}$$

the following equation results:

$$y_n = \frac{1}{\sum_{k=0}^{n} \alpha^k} \left(x_n + \left(\sum_{k=1}^{n} \alpha^k\right) y_{n-1}\right) = \frac{1-\alpha}{1-\alpha^n} \left(x_n + \frac{\alpha - \alpha^n}{1-\alpha} y_{n-1}\right) \tag{A.4.10}$$

This recursion yields an exponentially weighted moving average. However, due to the restrictions of the geometric sum, equation (A.4.10) is only valid for $\alpha<1$. Further, it can be interpreted as difference equation of a time-variant lowpass filter with a pole on the real axis. With increasing n, the pole approaches the value of α.

A.5 Ear canal variations

To test the accuracy of the determination method under variable conditions, a set of 12 different ear canal models was designed. The varied model parameters are canal length (long, medium, short), curvature (straight, curved) and inclination angle of the tympanic membrane (flat, steep). The models are depicted in Figure A 5.2. The results of the eardrum pressure estimations are shown in Fig. A 5.1.

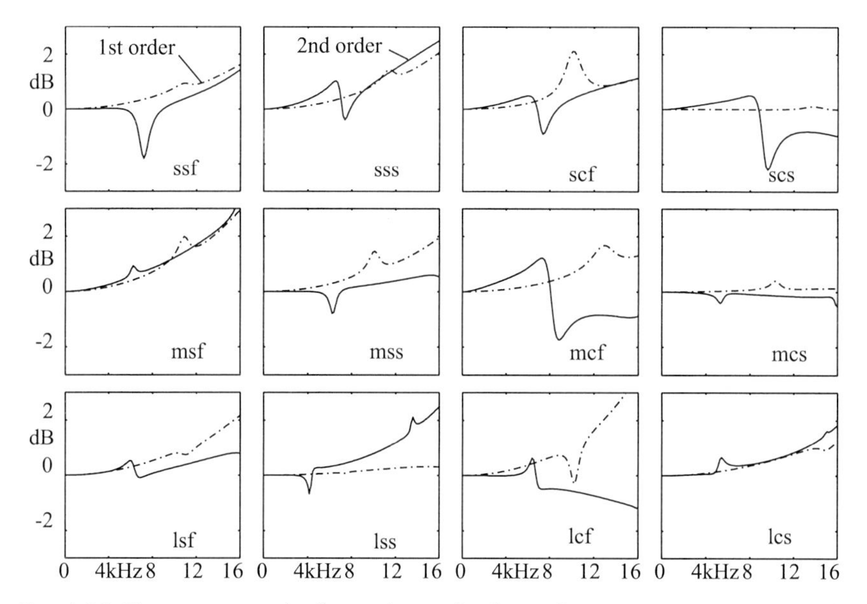

Fig. A.5.1: Estimation error for first and second order eardrum pressure estimations in the ear canals shown in Fig. A.5.2.The letter code in the panels indicates the canal length (long, medium, short), curvature (straight, curved) and inclination angle of the tympanic membrane (flat, steep).

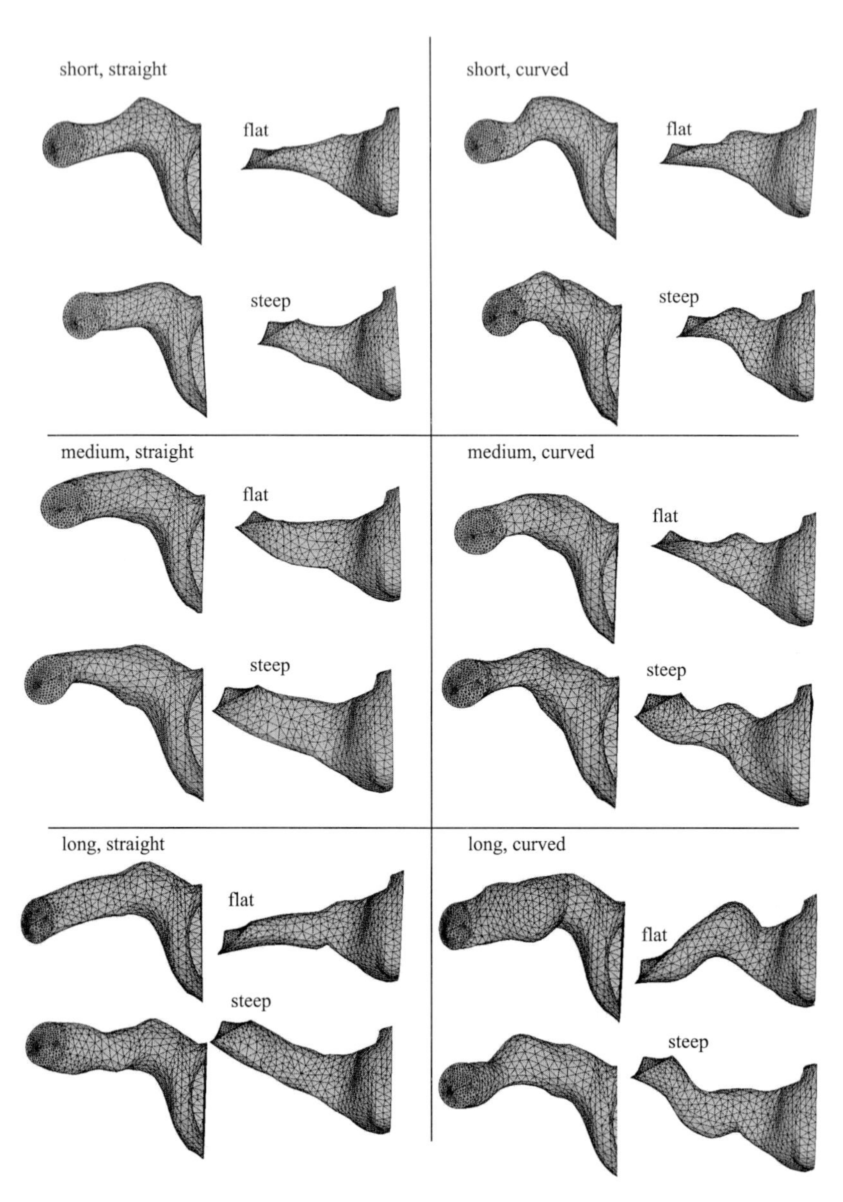

Fig. A.5.2: Geometry variations of the model ear canal.

References

Agullo, J., Barjau, A., and Keefe, D. H. (**1999**). "Acoustic Propagation in Flaring, Axisymmetric Horns: I. A New Family of Unidimensional Solutions," Acustica **85**, 278-284.

Barbone, P. E., and Gokhale, N. H. (**2004**). "Elastic modulus imaging: on the uniqueness and nonuniqueness of the elastography inverse problem in two dimensions," Inverse Problems **20**(1), 283-296.

Barjau, A., and Keefe, D. H. (**1999**). "Acoustic Propagation in Flaring, Axisymmetric Horns: II. Numerical Results, WKB Theory, and Viscothermal Effects," Acustica **85**, 285-293.

Bathe, K. J. (**1996**). Finite Element Procedures. Prentice-Hall, Englewood Cliffs.

Benade, A. H. (**1988**). "Equivalent circuits for conical waveguides," J. Acoust. Soc. Am. **83**(5), 1764-1769.

Benson, R. W. (**1953**). "The calibration and Use of Probe-Tube Microphones," J. Acoust. Soc. Am. **25**(1).

Blauert, J. (**1974**). Spatial Hearing. The psychophysics of human sound localization, Cambridge MA, MIT Press.

Böhme, J.F. (**1998**). Stochastische Signale,Teubner, Stuttgart.

Bornitz, M., Zahnert, T., Hardtke, H.-J., and Hüttenbrink, K.-B. (**1997**). Bestimmung von Materialparametern für elastischen Knorpel. Die Methode der Finiten Elemente in der Biomedizin und angrenzenden Gebieten, Ulm, Universität Ulm.

Botteldooren, D. (**1994**). "Acoustical finite-difference time-domain simulation in a quasi-Cartesian grid," J. Acoust. Soc. Am. **95**(5), 2313-2319.

Brand, T., and Hohmann, V. (**2002**). "An adaptive procedure for categorical loudness scaling," J. Acoust. Soc. Am. **112**(4), 1597-1604.

Brass, D., and Locke, A. (**1997**). "The effect of the evanescent wave upon acoustic measurements in the human ear canal," J. Acoust. Soc. Am. **101**(4), 2164-2175.

Brügel, F. J., Schorn, K., and Stecker, M. (**1990**). "In situ-Messungen zur Bestimmung der Wirkung von Zusatzbohrungen im Ohrpaßstück," Laryngo-Rhino-Otol. **69**, 337-340.

Bruhns, O., and Lehmann, T. (**1994**). Elemente der Mechanik II, Vieweg, Braunschweig.

Carlile, S., and Pralong, D. (**1994**). "The location-dependent nature of perceptually salient features of the human head-related transfer functions," J. Acoust. Soc. Am. **95**(6), 3445-3459.

Chan, J. C. K., and Geisler, C. D. (**1990**). "Estimation of eardrum acoustic pressure and of ear canal length from remote points in the canal," J. Acoust. Soc. Am. **87**(3), 1237-1247.

Chen, J., VanVeen, B. D., and Hecox, K. E. (**1992**). "External ear transfer function modeling: A beamforming approach," J. Acoust. Soc. Am. **92**(4), 1933-1944.

Ciric, D., and Hammershøi, D. (**2006**). "Coupling of earphones to human ears and to standard coupler," J. Acoust. Soc. Am. **120**(4), 2096-2107.

Ciric, D., and Hammershøi, D. (**2007**). "Acoustic impedances of ear canals measured by impedance tube," Proc. ICA 2007 (Madrid).

Curdes, Y., Hudde, H., and Taschke, H. (**2004**). “Modelling the Dynamic Behaviour of the Human Middle Ear Using the Finite-Element Method,” in: Proc. 7. Congrès Français d'Acoustique / 30. Dtsch. Jahrestg. f. Akust. (CFA/DAGA'04), F-Strassburg.

Dai, H. (**1995**). “On measuring psychometric functions: A comparison of the constant-stimulus and adaptive up-down methods,” J. Acoust. Soc. Am. **98**(6), 3135-3139.

Egolf, D. P. (**1977**). “Mathematical Modelling of a probe-tube Microphone,” J. Acoust. Soc. Am. **61**(1), 200-205.

Egolf, D. P., Feth, L. L., Cooper, W. A., and Franks, J. R. (**1985**). “Effects of normal and pathologic eardrum impedance on sound pressure in the aided ear canal: A computer simulation,” J. Acoust. Soc. Am. **78**(4), 1281-1285.

Egolf, D. P., Kennedy, W. A., and Larson, V. D. (**1992**). “Occluded-ear simulator with variable acoustic properties,” J. Acoust. Soc. Am. **91**(5), 2813-2823.

Egolf, D. P., Nelson, D. K., Howell, H. C., and Larson, V. D. (**1993**). “Quantifying ear-canal geometry with multiple computer-assisted tomographic scans,” J. Acoust. Soc. Am. **93**(5), 2809-2819.

Erkamp, R. Q., Skovoroda, A. R., Emelianov, S. Y., and O'Donnell, M. (**2004**). “Measuring the nonlinear elastic properties of tissue-like phantoms,” IEEE Trans Ultrason Ferroelectr Freq Control **51**(4).

Evans, N. W. (**1990**). “Superintegrability in classical mechanics,” Physical Review **41**(10), 5666-5676.

Farmer-Fedor, B. L., and Rabbitt, R. D. (**2002**). “Acoustic intensity, impedance and reflection coefficient in the human ear canal,” J. Acoust. Soc. Am. **112**(2), 600-620.

Fastl, H., Jaroszewski, A., Schorer, E., and Zwicker, E. (**1990**). “Equal Loudness Contours between 100 and 1000 Hz for 30,50 and 70 phon,” Acustica **70**, 197-201.

Feeney, M.P., and Sanford, C.A. (**2004**). "Age effects in the human middle ear: Wideband acoustical measures," J. Acoust Soc. Am. **116**, 3546-3558.

Fels, J. (**2008**). From children to Adults: How Binaural Cues and Ear Canal Impedances Grow. Doctoral Dissertation, Aachen RWTH.

Fels, J., Buthmann, P., and Vorländer, M. (**2004**). "Head-related transfer function of children," Acustica **90**(5), 918-927.

Fletcher, N. H., Smith, J., Tarnopolsky, A. Z., and Wolfe, J. (**2005**). "Acoustic impedance measurements-correction for probe geometry mismatch," J. Acoust. Soc. Am. **117**(5), 2889-2895.

Florentine, M., Buus, S., and Mason, C. R. (**1987**). "Level discrimination as a function of level for tones from 0.25 to 16 kHz," J. Acoust. Soc. Am. **81**(5), 1528-1541.

Funnell, W. R. J., and Decraemer, W. F. (**1996**). "On the incorporation of moiré shape measurements in finite-element models of the cat eardrum," J. Acoust. Soc. Am. **100**(2), 925-932.

Gabriel, B., Kollmeier, B., and Mellert, V. (**1994**). "Einfluß verschiedener Meßmethoden auf die Kurven gleicher Pegellautstärke," Fortschr. Akust. – DAGA'94, Dtsch. Ges. Akust., D-Oldenburg.

Gabriel, B., Kollmeier, B., and Mellert, V. (**1995**). "Kontexteffekte bei der Bestimmung der Kurven gleicher Pegellautstärke," Fortschr. Akust. – DAGA'95, Dtsch. Ges. Akust., D-Oldenburg.

Gabriel, B., Kollmeier, B., and Mellert, V. (**1997**). "Influence of Individual Listener, Measurement Room and Choice of Test-Tone Levels on the Shape of Equal-Loudness Level Contours," Acustica **83**, 670-683.

Gabriel, B., Mellert, V., and Kollmeier, B. (**1996**). "Modellierung von Kontexteffekten bei der Messung von Isophonen," Fortschr. Akust. – DAGA'96, Dtsch. Ges. Akust., D-Oldenburg.

Garner, W. R. (**1954**). "Context effects and the validity of loudness scales," J. Exp. Psychol. **48**(3).

Gilman, S., and Dirks, D. D. (**1986**). "Acoustics of ear canal measurement of eardrum SPL in simulators," J. Acoust. Soc. Am. **80**(3), 783-793.

Green, D. M. (**1993**). "A maximum-likelihood method for estimationg thresholds in a yes-no task," J. Acoust. Soc. Am. **93**(4), 2096-2104.

Gyo, K., Arimoto, H., and Goode, R. (**1987**). "Measurement of the ossicular vibration ratio in human temporal bones by use of a video measuring system," Acta Otolaryngol. **103**, 87-95.

Hammershøi, D., Hoffmann, P. F., Olesen, S. K., and Rubak, P. (**2008**). "Capturing blocked-entrance binaural signals from open-entrance recordings," Acoustics' 08: the 155th Acoustical Society of America Meeting, 5th Forum Acusticum, 9th Congrès Français d'Acoustique, June 29 to July 4, 2008, Paris, France.

Hammershøi, D., and Møller, H. (**1996**). "Sound transmission to and within the human ear canal," J. Acoust. Soc. Am. **100**(1), 408-427.

Hammershøj, D., and Møller, H. (**2008**). "Determination of Noise Immision From Sound Sources Close to the Ears," Acustica **94**, 114-129.

Hellbrück, J., and Ellermeier, W. (**2004**). Hören. Hogrefe-Verlag.

Hellstrom, P. A. (**1993**). "The relationship between sound transfer functions from free sound field to the eardrum and temporary threshold shift," J. Acoust. Soc. Am. **94**(3), 1301-1306.

Hellstrom, P. A., and Axelsson, A. (**1993**). "Miniature microphone probe tube measurements in the external auditory canal," J. Acoust. Soc. Am. **93**(2), 907-919.

Hoyer, H. E., and Dörheide, J. (**1983**). "A study of human head vibrations using time-averaged holography.," J Neurosurg **58**(5), 729-33.

Huang, G. T., Rosowski, J. J., Puria, S., and Peake, W. T. (**2000**). "A noninvasive method for estimating acoustic admittance at the tympanic membrane," J. Acoust. Soc. Am. **108**(3), 1128-1146.

Hudde, H. (**1983**). "Estimation of the area function of human ear canals by sound pressure measurements," J. Acoust. Soc. Am. **73**(1), 24-31.

Hudde, H. (**1983**). "Measurement of the eardrum impedance of human ears," J. Acoust. Soc. Am. **73**(1), 242-247.

Hudde, H. (**2005**). "A functional view on the peripheral human hearing organ," in: Communication Acoustics. Ed. J. Blauert, Springer, Berlin/Heidelberg/New York.

Hudde, H., and Engel, A. (**1998a**). "Acoustomechanical human middle ear properties. Part I: Model structure and measuring techniques," Acustica **84**, 720-738.

Hudde, H., and Engel, A. (**1998b**). "Acoustomechanical human middle ear properties. Part II: Ear canal middle ear cavities, eardrum, ossicles, and joints," Acustica **84**, 894-913.

Hudde, H., and Engel, A. (**1998c**). "Acoustomechanical human middle ear properties. Part III: Eardrum impedances, transfer functions, and model calculations," Acustica **84**, 1091-1108.

Hudde, H., and Engel, A. (**1999**). "Estimation of the sound pressure at eardrum," Joint Meeting ASA/EAA/DEGA, Forum Acusticum 1999, Acustica 85: Suppl. 1.

Hudde, H., Engel, A., and Lodwig, A. (**1999**). "Methods for estimating the sound pressure at the eardrum," J. Acoust. Soc. Am. **106**(4), 1977-1992.

Hudde, H., and Lackmann, R. (**1980**). "Systematische Fehler bei Messungen mit Sondenmikrofonen," Acustica **47**, 27-33.

Hudde, H., and Letens, U. (**1984**). "Messfehler bei Verwendung von Sondenmikrofonen," Fortschr. Akust. – DAGA'84, Dtsch. Ges. Akust., D-Oldenburg.

Hudde, H., Lodwig, A., and Engel, A. (**1996**). "A wide-band precision acoustic measuring head," Acustica **82**, 895-904.

Hudde, H., and Müller, M. (**2005**). "Einfluss der Gehörgangsform auf die Transformationseigenschaften des Gehörgangs," 31. Dtsch. Jahrestg. f. Akust. (DAGA'05), D-München.

Hudde, H. and Schmidt, S. (**2009**). "Sound fields in generally shaped curved ear canals", J. Acoust. Soc. Am. 125(5), in print.

Hughes, T.J.R. (**2000**). The Finite Element Method: Linear Static and Dynamic Finite Element Analysis. Dover Publications.

IEC (**1986**). "Occluded-ear simulator for the measurement of earphones coupled to the ear by ear inserts," IEC711.

ISO (**1999**). "Acoustics – Reference zero for the calibration of audiometric equipment," EN ISO 389-1.

ISO (**2006**). "Acoustics - Normal equal-loudness contours," ISO/DIS 226:2001.

ITU-T (**2002**). "Telephone transmission quality, telephone installations, local line networks – Objective measuring apparatus," ITU-T P.57.

Karal, F. C. (**1953**). "The Analogous Acoustical Impedance for Discontinuities and Constrictions of Circular Cross Section," J. Acoust. Soc. Am. **25**(2), 327-334.

Katz, B. F. G. (**2000**). "Acoustic absorption measurement of human hair and skin within the audible frequency range," J. Acoust. Soc. Am. **108**(5), 2238-2242.

Katz, B. F. G. (**2000**). "Method to resolve microphone and sample location errors in the two-microphone duct measurement method," J. Acoust. Soc. Am. **108**(5), 2231-2237.

Katz, B. F. G. (**2001**). "Boundary element method calculation of individual head-related transfer function. I. Rigid model calculation," J. Acoust. Soc. Am. **110**(5), 2440-2455.

Katz, B. F. G. (**2001**). "Boundary element method calculation of individual head-related transfer function. II. Impedance effects and comparisons to real measurements," J. Acoust. Soc. Am. **110**(5), 2440-2455.

Keefe, D. H., Bulen, J. C., Campbell, S. L., and Burns, E. M. (**1994**). "Pressure transfer function and absorption cross section from the diffuse field to the human infant ear canal," J. Acoust. Soc. Am. **95**(1), 355-371.

Keefe, D. H., Bulen, J. C., Hoberg Arehart, K. and Burns, E. M. (**1993**). "Ear-canal impedance and reflection coefficient in human infants and adults," J. Acoust. Soc. Am. **94**(5), 2167-2638.

Keefe, D. H., Ling, R. and Bulen, J. C. (**1992**). "Method to measure acoustic impedance and reflection coefficient," J. Acoust. Soc. Am. **91**(1), 470-485.

Khanna, S. M., and Stinson, M. R. (**1985**). "Specification of the acoustical input to the ear at high frequencies," J. Acoust. Soc. Am. 77(2), 577-589.

Killion, M. C. (**1978**). "Revised estimate of minimum audible pressure: Where is the "missing 6 dB"?," J. Acoust. Soc. Am. **63**(5), 1501-1508.

Kringlebotn, M., and Gundersen, T. (**1985**). "Frequency characteristics of the middle ear," J. Acoust. Soc. Am. **77**, 159-164.

Kulik, Y. (**2007**). "Transfer matrix of conical waveguides with any geometric parameters for increased precision in computer modeling," J. Acoust. Soc. Am. **122**(5).

Lanoye, R., Vermeir, G., and Lauriks, W. (**2006**). "Measuring the free field acoustic impedance and absorption coefficient of sound absorbing materials with a combined particle velocity-pressure sensor," J. Acoust. Soc. Am. **119**(5), 2826-2831.

Larson, V. D., Nelson, J. A., Cooper, W. A., and Egolf, D. P. (**1993**). "Measurement of acoustic impedance at the input to the occluded ear canal," Journal of Rehabilitation Research and Development **30**(1), 129-136.

Lawrenson, C. C., Lafleur, L. D., and Shields, F. D. (**1998**). "The solution for the propagation of sound in a toroidal waveguide with driven walls (the acoustitron)," J. Acoust. Soc. Am. **103**(3), 1253-1260.

Libby, E. R., and Westermann, S. (**1988**). Principles of acoustic measurement and ear canal resonances. Handbook of Hearing Aid Amplification, Volume I and II. R. E. Sandlin, College Hill Press.

Lodwig, A., and Hudde, H. (**1998**). "Trommelfellbezogene Audiometrie," Fortschr. Akust. – DAGA'98, Dtsch. Ges. Akust., D-Oldenburg.

Lopez-Poveda, E. A., and Meddis, R. (**1996**). "A physical model of sound diffraction and reflections in the human concha," J. Acoust. Soc. Am. **100**(5), 3248-3259.

Mehrgardt, S., and Mellert, V. (**1977**). "Transformation characteristics of the external human ear," J. Acoust. Soc. Am. **61**(6), 1567-1576.

Møller, A. R. (**1960**). "Improved Technique for Detailed Measurements of the Middle Ear Impedance," J. Acoust. Soc. Am. **32**(2), 250-257.

Møller, H., Hammershøi, D., Jensen, C. B., and Sørensen, M. F. (**1995**). "Transfer Characteristics of Headphones Measured on Human Ears," J. Audio Eng. Soc. **43**(4), 203-217.

Müller, J., Oswald, J. A., and Janssen, T. (**2004**). "Probleme bei der Kalibrierung von Ohrsonden zur Messung von Distorsionsprodukten otoakustischer Emissionen," Z. Audiol. **43**(3), 112-123.

Neely, S. T., and Gorga, M. P. (**1998**). "Comparison between intensity and pressure as measures of sound level in the ear canal," J. Acoust. Soc. Am. **104**(5), 2925-2934.

Nordahn, M. (**2005**). “Modeling individual ear canal geometries and the effect of ventilation in hearing aids,” 31. Dtsch. Jahrestg. f. Akust., DAGA '05, D-München.

Ophir, J., Céspedes, I., Ponnekanti, H., Yazdi, Y., and Li, X. (**1991**). “Elastography: a quantitative method for imaging the elasticity of biological tissues,” Ultrasonic Imaging **13**, 111-134.

Oswald, J. A. (**2005**). Objektive Audiometrie mit otoakustischen Emissionen und akustisch evozierten Potenzialen. Doctoral Dissertation, TU Munich, 2005.

Oswald, J. A., Rosner, T., and Janssen, T. (**2006**). “Hybrid measurement of auditory steady-state responses and distortion product otoacoustic emissions using an amplitude-modulated primary tone,“ J. Acoust. Soc. Am. **119**(6), 3886–3895.

Parker, J.R. (1997). Algorithms for Image Processing and Computer Vision, New York, John Wiley & Sons, Inc.

Petyt, M. (1990). Introduction to Finite Element Vibration Analysis. Cambridge University Press.

Pralong, D., and Carlile, S. (**1994**). “Measuring the human head-related transfer functions: A novel method for the construction and calibration of a miniature “in-ear” recording system,” J. Acoust. Soc. Am. **95**(6), 3435-3444.

Pralong, D., and Carlile, S. (**1996**). “The role of individualized headphone calibration for the generation of high fidelity virtual auditory space,” J. Acoust. Soc. Am. **100**(6), 3785-3793.

Puria, S., Peake, W. T., and Rosowski, J. J. (**1997**). “Sound-pressure measurements in the cochlear vestibule of human-cadaver ears,” J. Acoust. Soc. Am. **101**(5), 2754-2770.

Qi, L., Funnell, W. R. J., and Daniel, S. J. (**2008**). “A nonlinear finite-element model of the newborn middle ear,” J. Acoust. Soc. Am. **124**(1), 337-347.

Rabbitt, R. D. (**1990**). “A hierarchy of examples illustrating the acoustic coupling of the eardrum,” J. Acoust. Soc. Am. **87**(6), 2566-2582.

Rabbitt, R. D., and Friedrich, M. T. (**1991**). "Ear canal cross-sectional pressure distributions: Mathematical analysis and computation," J. Acoust. Soc. Am. **89**(5), 2379-2390.

Rabbitt, R. D., and Holmes, M. H. (**1988**). "Three-dimensional acoustic waves in the ear canal and their interaction with the tympanic membrane," J. Acoust. Soc. Am. **83**(3), 1064-1080.

Rabinowitz, W. M. (**1981**). "Measurement of the acoustic input immittance of the human ear," J. Acoust. Soc. Am. **70**(4), 1025-1035.

Reckhardt, C. (**2000**). Factors influencing equal-loudness level contours. Doctoral Dissertation, Oldenburg.

Reckhardt, C., Mellert, V., and Kollmeier, B. (**1998**). "Bestimmung von Isophonen mit einem Adaptiven Verfahren - Einfluß experimenteller Parameter auf die Ergebnisse," Fortschr. Akust. – DAGA'98, Dtsch. Ges. Akust., D-Oldenburg.

Rostafinski, W. (**1972**). "On propagation of long waves in curved ducts," J. Acoust. Soc. Am. **52**, 1411-1420.

Sachs, R. M., and Burkhard, M. D. (**1972**). "Insert earphone pressure response in real ears and couplers," J. Acoust. Soc. Am. **52**, 183.

Sanborn, P.-E. (**1998**). "Predicting hearing aid response in real ears," J. Acoust. Soc. Am. **103**(6), 3407-3417.

Scheperle, R. A., Neely, S. T., Kopun, J. G., and Gorga, M. P. (**2008**). "Influence of *in-situ*, sound-level calibration on distortion-product otoacoustic emission variability," J. Acoust. Soc. Am. **124**, 288-300.

Schmidt, R., Volkmer, T., Wendlang, G. and Buchelt, B. (**2002**). Effective elasticity constants of biomaterial presented on wood structure. FEM Workshop 2002 - The Finite Element Method in Biomedical Engineering, Biomechanics and Related Fields, Ulm, Universität Ulm.

Schmidt, S., and Hudde, H. (**2004**). "Finite Difference Time Domain Simulation of the Outer Ear." Proc. 7. Congrès Français d'Acoustique / 30. Dtsch. Jahrestg. f. Akust. (CFA/DAGA'04), F-Strasbourg.

Schmidt, S., and Hudde, H. (**2009**). "Accuracy of acoustic impedances measured at the ear canal entrance," J. Acoust. Soc. Am., in print (April 2009).

Schorn, K., and Stecker, M. (**1994**). Hörprüfungen. Oto-Rhino-Laryngologie in Klinik und Praxis. J. H. H.H. Naumann, K. Herberhold, E. Kasterbauer. Stuttgart, Thieme Verlag. **1:** 309-368.

Schroeder, M. (**1967**). "Determination of the Geometry of the Human Vocal Tract by Acoustic Measurements," J. Acoust. Soc. Am. **41**(4), 1002-1010.

Shaw, E. A. G., and Teranishi, R. (**1968**). "Sound Pressure Generated in an External-Ear Replica and Real Human Ears by a Nearby Point Source," J. Acoust. Soc. Am. **44**(1), 240-249.

Siegel, J. H. (**1994**). "Ear-canal standing waves and high-frequency sound calibration using otoacoustic emission probes," J. Acoust. Soc. Am. **95**(5), 2589-2597.

Sobotta, J. (**2000**). Atlas der Anatomie des Menschen, Band 1: Kopf, Hals, obere Extremität. Urban und Fischer, 21st edition.

Sondhi, M. M., and Gopinath, B. (**1971**). "Determination of Vocal-Tract Shape from Impulse Response at the Lips," J. Acoust. Soc. Am. **49**(6), 1867-1873.

Stevens, K. N., Berkovitz, R., Kidd, G., and Green, D. M. (**1987**). "Calibration of ear canals for audiometry at high frequencies," J. Acoust. Soc. Am. **81**(2), 470-484.

Stinson, M. R. (**1985a**). "The spatial distribution of sound pressure within scaled replicas of the human ear canal," J. Acoust. Soc. Am. **78**(5), 1596-1602.

Stinson, M. R. (**1985b**). "Spatial variation of phase in ducts and the measurement of acoustic energy reflection coefficients," J. Acoust. Soc. Am. 77(2), 386-393.

Stinson, M. R. (**1990**). “Revision of estimates of acoustic energy reflectance at the human ear-drum,” J. Acoust. Soc. Am. **88**(4), 1773-1778.

Stinson, M. R. (**1991**). “The propagation of plane sound waves in narrow and wide circular tubes, and generalization to uniform tubes of arbitrary cross-sectional shape,” J. Acoust. Soc. Am. **89**(2), 550-558.

Stinson, M. R., and Daigle, G. A. (**2005**). “Comparison of an analytic horn equation approach and a boundary element method for the calculation of sound fields in the human ear canal,” J. Acoust. Soc. Am. **118**(4), 2405-2411.

Stinson, M. R., and Daigle, G. A. (**2007**). “Transverse pressure distributions in a simple model ear canal occluded by a hearing aid test fixture,” J. Acoust. Soc. Am. **121**(6), 3689-3702.

Stinson, M. R., and Daigle, G. A. (**2008**). “The sound fi eld in life-size replicas of human ear canals occluded by a hearing aid,” Acoustics' 08: the 155th Acoustical Society of America Meeting, 5th Forum Acusticum, 9th Congrès Français d'Acoustique, June 29 to July 4, 2008, Paris, France.

Stinson, M. R., and Khanna, S. M. (**1989**). “Sound propagation in the ear canal and coupling to the eardrum, with measurements on model systems,” J. Acoust. Soc. Am. **85**(6), 2481-2491.

Stinson, M. R., and Khanna, S. M. (**1994**). “Spatial distribution of sound pressure and energy flow in the ear canals of cats,” J. Acoust. Soc. Am. **96**(1), 170-180.

Stinson, M. R., and Lawton, B. W. (**1989**). “Specification of the geometry of the human ear canal for the prediction of sound-pressure level distribution,” J. Acoust. Soc. Am. **85**(6), 2492.

Stinson, M. R., Shaw, E. A. G., and Lawton, B. W. (**1982**). “Estimation of acoustical energy reflectance at the eardrum from measurements of pressure distribution in the human ear canal,” J. Acoust. Soc. Am. **72**, 766-773.

Storey, L., and Dillon, H. (**2001**). "Estimating the location of probe microphones relative to the tympanic membrane," J. Am. Acad. Audiol. **12**(150-154).

Strube, H. W. (**2003**). "Are conical segments useful for vocal-tract simulation?," J. Acoust. Soc. Am. **114**(6), 3028-3031.

Suzuki, Y., Mellert, V., Richter, U., Møller, H., Nielsen, L., Hellman, R., Ashihara, K., Ozawa, K., and Takeshima, H. (**2003**). "Precise and full-range determination of two-dimensional equal loudness contours," NEDO project for ISO revision (E).

Suzuki, Y., and Takeshima, H. (**2004**). "Equal-loudness-level contours for pure tones," J. Acoust. Soc. Am. **116**(2), 918-933.

Takeshima, H., Suzuki, Y., Fujii, H., Kumagai, M., Ashihara, K., Fujimori, T., and Sone, T. (**2001**). "Equal-Loudness Contours Measured by the Randomized Maximum Likelihood Sequential Procedure," Acustica **87**, 389-399.

Taschke, H. (**2005**). Mechanismen der Knochenschallleitung. Doctoral Dissertation, Ruhr-Universität Bochum.

Taschke, H., and Hudde, H. (**2006**). "A finite element model of the human head for auditory bone conduction simulation," J. Oto-Rhino-Laryngology **68**, 324-328.

Tonndorf, J., and Khanna, S. M. (**1972**). "Tympanic-Membrane Vibrations in Human Cadaver Ears Studied by Time-Averaged Holography," J. Acoust. Soc. Am. **52**(4), 1221-1233.

Tuck-Lee, J. P., Pinsky, P. M., Steele, Ch. R., and Puria, S. (**2008**). "Finite-element modeling of acousto-mechanical coupling in the cat middle ear," J. Acoust. Soc. Am. **124**(1), 348-362.

Ulrich, R., and Miller, J. (**2004**). "Threshold estimation in two-alternative forced-choice (2AFC) tasks: The Spearman-Kärber method," Perception & Psychophysics **66**(3), 517-533.

Vallejo, L. A., Delgado, V. M., Hidalgo, A., Gil-Carcedo, E., Gil-Carcedo, L. M., and Montoya, F. (**2006**). "Modelado de la geometría del conducto auditivo externo mediante el método de los elementos finitos," Acta Otorrinolaringol Esp **57**, 82-89.

Verhey, J., and Kollmeier, B. (**1998**). "Messungen zur zeitabhängigen Lautheitssummation," Fortschr. Akust. – DAGA'98, Dtsch. Ges. Akust., D-Oldenburg.

von Bekesy, G. (**1948**). "Vibration of the head in a Sound Field and Its Role in Hearing by Bone Conduction," J. Acoust. Soc. Am. **20**(6).

Vorländer, M. (**2000**). "Acoustic load on the ear caused by headphones," J. Acoust. Soc. Am. **107**(4), 2082-2088.

Voss, S. E., and Allen, J. B. (**1994**). "Measurement of acoustic impedance and reflectance in the human ear canal," J. Acoust. Soc. Am. **95**(1), 372-384.

Voss, S. E., Rosowski, J. J., Shera, C. A., and Peake, W. T. (**2000**). "Acoustic mechanisms that determine the ear-canal sound pressures generated by earphones," J. Acoust. Soc. Am. **107**(3), 1548-1565.

Wada, H., Koike, T., and Kobayashi, T. (**2002**). "Modeling of the human middle ear using the finite-element method," J. Acoust. Soc. Am. **111**(3), 1306-17.

Webster, A. G. (**1919**). "Acoustical impedance, and the theory of horns and of the phonograph," Proc. Natl. Acad. Sci. U.S.A. **5**, 275-282.

Weistenhöfer, C., and Hudde, H. (**1999**). "Determination of the shape and inertia properties of the human auditory ossicles," Audiology and Neuro-Otology **4**, 192-196.

Weistenhöfer, C. (**2002**). Funktionale Analyse des menschlichen Mittelohres durch dreidimensionale Messung und Modellierung. Doctoral Dissertation, Ruhr-Universität Bochum.

Whitehead, M. L., Simons, I., Stagner, B. B., and Martin, G. K. (**1997**). "The frequency response of the ER-2 speaker at the eardrum," J. Acoust. Soc. Am. **101**(2), 1195-1198.

Whitehead, M. L., Stagner, B. B., Lonsbury-Martin, B. L., and Martin, G. K. (**1995**). "Effects of ear-canal standing waves on measurements of distorsion-product otoacoustic emissions," J. Acoust. Soc. Am. **98**(6), 3200-3213.

Zannoni, C., Mantovani, R., and Viceconti, M. (**1998**). "Material properties assignment to finite element models of bone structures: a new method," Medical Engineering & Physics **20**, 735-740.

Zienkiewicz, O. C., and Newton, R. E. (**1969**). "Coupled Vibrations of a Structure Submerged in a Compressible Fluid," Proceedings of the Symposium on Finite Element Techniques, University of Stuttgart, Germany.

Zienkiewicz, O. C., and Taylor, R.L. (**2000**). The Finite Element Method, Fifth Edition, Volume 3: Fluid Dynamics. Butterworth-Heinemann, Oxford.

Zollner, M., and Zwicker, E. (**1993**). Elektroakustik. Springer, Berlin/Heidelberg/New York.

Zwicker, E., and Fastl, H. (**1999**). Psychoacoustics. Springer, Berlin/Heidelberg/New York.

Curriculum Vitae

Name:	Sebastian Schmidt
Date and place of birth:	07.07.1977, Bochum-Langendreer (Germany)

Education

08/1983 – 06/1985	Primary School, "Städtische Gemeinschaftsgrundschule Alte Bahnhofstraße", Bochum
08/1985 – 06/1987	Primary School, "Gemeinschaftsgrundschule Kemnader Straße", Bochum
08/1987 – 08/1996	Grammar School (Schiller-Gymnasium), Bochum General qualification for university entrance: June 1996

Alternative service

09/1996 – 09/1997	Retirement home "Lutherheim", Bochum

Studies

10/1997 – 09/1998	Study of physics and educational science at the Ruhr-University Bochum
10/1998 – 12/2003	Study of electrical engineering at the Ruhr-University Bochum, Diploma degree in electrical engineering (Dipl.-Ing.)
01/2004 – 06/2009	Ph.D. student and employment as research assistant at the Auditory Acoustics Research Group (Institute of Communication Acoustics, faculty of Electrical Engineering and Computer Science) at the Ruhr-University Bochum

Danksagung

Die Entstehung dieser Arbeit hing von der Unterstützung vieler Menschen ab, bei denen ich mich an dieser Stelle herzlich bedanken möchte. Ein erster Dank geht an Prof. Dr.-Ing. Herbert Hudde, der mir die Möglichkeit gab, dieses äußerst interessante Projekt zu bearbeiten und als mein Doktorvater die wissenschaftliche Betreuung übernahm. In seiner Forschungsgruppe Hörakustik habe ich in meinen „akustischen Lehrjahren" eine höchst spannende und und durch eine äußerst offene Atmosphäre fachlicher Diskussion geprägte Zeit verbracht.

Für die Übernahme des Zweitgutachtens und das wohlwollende Interesse, das er dieser Arbeit entgegenbrachte, bedanke mich bei Prof. Dr. rer. nat. Michael Vorländer.

Von Dipl.-Ing. Yvonne Curdes und Dr.-Ing. Henning Taschke, meinen ehemaligen Kollegen in der Forschungsgruppe, habe ich den praktischen Umgang mit Finiten Elementen gelernt. Da dieser für die vorliegende Arbeit von entscheidender Bedeutung war, gilt ihnen dafür ein besonderer Dank, jedoch noch mehr für die freundschaftliche Aufnahme und das tolle Gruppenklima.

Am Institut für Kommunikationsakustik konnte ich stets in einer sehr angenehmen Arbeitsatmosphäre forschen. Dafür danke ich dem Institutsleiter Prof. Dr.-Ing. Rainer Martin und meinen ehemaligen wissenschaftlichen Kollegen Dipl.-Ing. Christian Borß, Dr.-Ing. Colin Breithaupt, Dr.-Ing. Gerald Enzner, Dipl.-Ing. Sebastian Gergen, Dipl.-Ing. Timo Gerkmann, Dr.-Ing. Nilesh Madhu, Sarmad Malik M.Sc., Dipl.-Ing. Dirk Mauler, Dipl.-Ing. Anil Nagathil, Dipl.-Ing. Alexander Schasse und Dipl.-Ing. Dominic Schmid.

Für mehr als nur administrative oder technische Hilfestellung sei den nichtwissenschaftlichen Kollegen des IKA gedankt: Ewa Figiel, Andreas Hinz, Edith Klaus, Wolfgang Kloiber,

Renate Leopold, Felipe Pareja-Reina und Peter Salzsieder (Diskussionspartner für Audiohardwarefragen aller Art, sei es die Speisespannungsversorgung der Sondenmikrofone, seien es Röhrenendstufen). Dank sei auch der mechanischen Werkstatt für die gute Zusammenarbeit bei der „Hardwareausführung“ gesagt, vor allem Hans-Joachim Fischer, Herbert Kellner, Werner Staal.

Selbstverständlich wurde das Projekt durch Diplomanden, Studienarbeiter und studentische Hilfskräfte unterstützt. Mein Dank geht an Dipl.-Ing. Mohammed Abdolhamidi, cand.-Ing. Sebastian Becker, Dipl.-Ing. Sven Bergmann, cand.-Ing. Nils Diemel, Dipl.-Ing. Sven Dyrbusch, cand.-Ing. Johannes Gauer, Dipl.-Ing. Karthika Gnanasegaram, cand.-Ing. Raphael Koning, Dipl.-Ing. Michał Mleczko.

Die an den Experimenten teilnehmenden Versuchspersonen haben in besonderer Weise zu diesem Projekt beigetragen. Dafür danke ich Sabina Barczak, Nils Diemel, Leona Ehlert, Andreas Gaidt, Raphael Koning, Marco Pech und vor allem Frank Wellner, dessen Pinna man in dieser Arbeit in vielen Illustrationen findet.

Sehr hohe Trommelfellschalldrücke $p_{T,max}$ konnten immer dann nachgewiesen werden, wenn die „Gehör-Gang“ zur Weihnachtsfeier spielte, also die aus Institutsmitarbeitern und -studierenden bestehende Band (A. Schasse, A. Nagathil, R. Koning, S. Dyrbusch, P. Salzsieder, M. Mleczko, T. Gerkmann, J. Gauer und der Autor dieses Textes).

Schließlich bedanke ich mich bei meinen Eltern, ohne deren Unterstützung mein Studium und somit auch diese Arbeit nicht möglich gewesen wäre. Ein herzlicher Dank für Rückhalt und Verständnis gilt auch meinem Bruder und meiner übrigen Familie sowie meinen Schwiegereltern. Vor allem danke ich jedoch meiner Frau Verena für ihr Vertrauen und ihre Liebe und werde versuchen, ihr dies bei ihrer eigenen, nun anstehenden Dissertationsprüfung genau so zu vergelten.